The Spacefarer's Handbook

Bergita Ganse • Urs Ganse

The Spacefarer's Handbook

Science and Life Beyond Earth

 Springer

Bergita Ganse
Manchester Metropolitan University
Manchester, UK

Urs Ganse
University of Helsinki
Helsinki, Finland

SPRINGER-PRAXIS BOOKS IN SPACE EXPLORATION

Springer Praxis Books
Space Exploration
ISBN 978-3-662-61701-4   ISBN 978-3-662-61702-1   (eBook)
https://doi.org/10.1007/978-3-662-61702-1

Original German edition published by Springer, 2017

Cover Figure: iss061e143462 (Jan. 25, 2020) — ESA (European Space Agency) astronaut Luca Parmitano is pictured tethered to the International Space Station while finalizing thermal repairs on the Alpha Magnetic Spectrometer, a dark matter and antimatter detector, during a spacewalk that lasted 6 hours and 16 minutes. Credit: NASA. Source: https://www.nasa.gov/image-feature/esa-astronaut-luca-parmitano-is-tethered-to-the-space-station-0.

This Springer imprint is published by the registered company Springer-Verlag GmbH, DE, part of Springer Nature.
The registered company address is: Heidelberger Platz 3, 14197 Berlin, Germany

Preface

Born as a spacefarer and just not been to space yet? Getting ready for a flight? Currently on orbit and got a question? This book explains the practical reality of space travel and covers the most important fields—from spaceship construction, planning and navigation via life in space and medicine in weightlessness to exploration and the search for extraterrestrial life. It is a guidebook written by scientists!

As an astrophysicist and a space physician, we professionally cover entirely different areas of the field. We are siblings and have spent a significant amount of our childhood enjoying science fiction (in particular Star Trek), stargazing, simulating flying spacecraft and colonising planets on various computer systems and consoles. We are excited to be a part of the space age and look forward to the further exploration of our solar system and beyond. We hope to share and celebrate the atmosphere of fascination that we have experienced and still feel every single day while working as scientists in two separate areas of space research. As such, we feel honoured to be a part of this great adventure!

Despite human spaceflight being a gigantic interdisciplinary and international masterpiece, it all comes down to basic science. In our book, we want to unveil some of the spaceflight magic and correct the belief that only superheroes can fly to space, as well as the misconception that spacecraft only work when an alien artefact from the future grants its power. This book presents interesting facts, phenomena, anecdotes and tips. We frequently laughed our heads off when finding new stories and fun facts. In addition, we are in contact with a number of people who have been to space and who inspired us and answered our questions to make this book a compendium of real-life spaceflight knowledge. We are especially thankful for astronaut Story

Musgrave's permission to print an interview with him. As he so eloquently put it, "I don't collect the data, I am the data!".

Many aspiring spacefarers are eager to find out how they can make their own spaceflight a reality. For others, the matter is already settled or the flight currently takes place. Someone may find the idea ridiculous to ever enter a spacecraft and feel safe and comfortable on the ground. This book is intended for all reader groups! We want to help everyone in successfully planning and undertaking their flight. In addition, we hope to feed those readers who think they know it all with some exciting and enjoyable additional facts.

In 2017, we published the Handbook for Aspiring Spacefarers in German. Since our non-German-speaking friends kept complaining they wanted an English version, we have now translated, updated and extended our book and hope it turned out to be an enjoyable read!

Have a good flight!

Manchester, UK   Bergita Ganse
Helsinki, Finland   Urs Ganse

Additional information, videos, downloads and other links are sometimes included as QR-codes. A QR-code scanner (app) may be used to access the content, or the URL be typed into a web browser.

Acknowledgements

This book would not have been possible without the help, patience and support of our colleagues, friends and family over several months.

We would especially like to thank the proofreaders, Julia Attias, Harriet George, Alex Ireland, Diana Morosan and Marje Niemelä for bearing with our sometimes puzzling use of the English language.

Lisa Edelhäuser and Ramon Khanna of Springer deserve a special mention for supporting us in the authoring and publishing process.

Contents

About the Authors

PD Dr. Bergita Ganse is an Orthopaedic Surgeon and a Physiologist with a research focus on Space Medicine and the musculoskeletal system in spaceflight. She has received her Dr. med. from Lübeck University and her habilitation (postdoctoral thesis) from RWTH Aachen University in Germany. She is currently a Research Fellow at Manchester Metropolitan University in the UK, funded by the German Research Foundation (DFG). She is a co-investigator of an ISS experiment and involved in large international studies working with the German Aerospace Center (DLR), the European Space Agency (ESA) and the National Aeronautics and Space Administration (NASA).

Dr. Urs Ganse is a Theoretical Space Physicist, with a research focus on plasma simulations. After studying physics and obtaining his doctorate from the University of Würzburg, Germany, on the subject of solar radio bursts, he has worked as a postdoc in Finland and South Africa, with research funding from DFG, ESA and the Academy of Finland. In his current position as a University Researcher at the University of Helsinki, he uses supercomputers to model the near-Earth plasma environment and its interactions with Earth's magnetic field.

1

How to Become a Spacefarer

Contents

1.1   Nothing Ventured, Nothing Gained

Why are the most capable spacefarers usually selected from thousands of applicants? Because space agencies can choose from big crowds of volunteers desperately wanting to go! The selection processes limit risks associated with the "human factor", which includes health and psychology, and also ensures candidates are skilled in a multitude of fields. So, there is actually no real reason that makes most applicants ineligible. In fact, most humans would be totally able to survive a spaceflight in good order. As most of those who would have loved to go have not made it to space so far, and as job offers for spacefarers are rare, we have collected potential options to grab a flight opportunity:

1. Applying for a job advertisement—for example, when one of the big space agencies searches for new candidate spacefarers. The European Space Agency (ESA) last selected six candidates in the year 2008. In 2016, women

© Springer-Verlag GmbH Germany, part of Springer Nature 2020
B. Ganse, U. Ganse, *The Spacefarer's Handbook*, Springer Praxis Books,
https://doi.org/10.1007/978-3-662-61702-1_1

in Germany could apply to become the first German woman in space. The best odds seem to exist in the USA, where astronaut candidates are annually selected by the National Aeronautics and Space Administration (NASA). Other nations with a human spaceflight programme include China, Russia, Canada and Japan.

2. Buying a ticket. Many companies have sold tickets in the past, but most of them were unable to deliver! All space tourists until now have spent multiple days on the International Space Station (ISS). It has been the only destination for space tourists up until now. However, shorter flights should also be available soon.

3. Spending a big sum of money to build an own spacecraft or to buy one from a commercial company. This option has previously only been successfully applied by the company Scaled Composites—however with funding by Microsoft founder and billionaire Paul Allen. Although one could argue that it does not count as a do-it-yourself project when an established aerospace company funded by an IT-billionaire builds a spacecraft.

4. (The strategy to wait until aliens come to pick one up is probably less promising…)

By now, humans from many countries have been to space. The Soviet Union/Russia, the USA and China have been the only countries to build and launch their own spacecraft carrying humans into space. Many other nations have used their infrastructure to bring their own spacefarers into orbit. Depending on the country that has built the vehicle, these spacefarers are called either *astronauts*, *cosmonauts* or *taikonauts*. One is an astronaut when the rocket was American, a cosmonaut if one went with the Russians or previously with the Soviet Union, and a taikonaut if one had the chance to fly with the Chinese (the ending "naut" meaning sailor). Despite this, India calls their spacefarers *vyomanauts*, Malaysia theirs *angkasawan* and Germany *raumfahrer*. Figure 1.1 shows a world map indicating all launch sites humans have departed from so far.

1.2   Open Challenges and the Story So Far

Figure 1.2 is an illustration of the most important first time events of human spaceflight's pioneering days—as a refresher, inspiration and motivation to add new achievements, such as the self-experienced first landing on an asteroid or a planet.

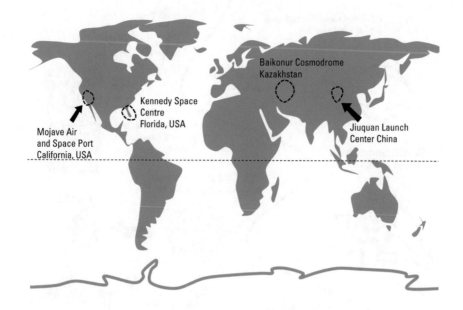

Fig. 1.1 Map of the world showing all launch sites that have been used for crewed spaceflight so far. The biggest spaceport of the world (by area) is Baikonur Cosmodrome in Kazakhstan. All crewed NASA missions including all Space Shuttle flights have departed from the Kennedy Space Center in Florida. The only exception is the SpaceShipOne, that launched from California. The Chinese crewed rockets depart from the Jiuquan Launch Centre

So far, the Soviet Union/Russia, China and the USA have had their own spacestations in Earth's orbit. The Soviet Union's Salyut 1 station was the first spacestation in human history and had numerous successors, including the Almaz military spacestations. These spacestations were designated as Salyut stations to hide their military purpose (all three Almaz spacestations were armed with an anti-air cannon). Overall, seven Salyut stations were delivered to orbit. Salyut 2 was the only one to lose altitude and depressurise, which occurred after only 2 weeks. The other Salyut stations were operational, and Salyut 7 even stayed in orbit for almost 9 years until 1991. The last spacestation of the Soviet Union and later Russia was Mir (1986–2001), an amazing success story with many visits from international spacefarers and American Space Shuttles. The name Mir is a pun, and means *peace* and *world* at the same time. On the US-side, the huge spacestation Skylab was operational from 1973 to 1979. With 360 m³, this station was bigger than Mir (350 m³). It

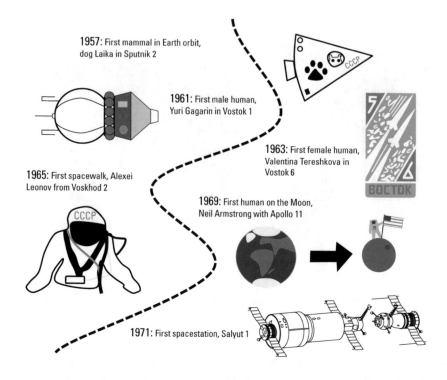

1957: First mammal in Earth orbit, dog Laika in Sputnik 2

1961: First male human, Yuri Gagarin in Vostok 1

1963: First female human, Valentina Tereshkova in Vostok 6

1965: First spacewalk, Alexei Leonov from Voskhod 2

1969: First human on the Moon, Neil Armstrong with Apollo 11

1971: First spacestation, Salyut 1

Fig. 1.2 First times in the history of crewed spaceflight

was, however, the only operational and crewed US spacestation in history. The planned spacestation *Freedom* was never built. Instead, parts and concepts of it were later used to create the International Space Station, which is an international endeavour. Russia's contribution to the ISS originated in the same way from a planned and never built spacestation *Mir 2*. China had the crewed spacestations Tiangong-1 (2011–2018) and Tiangong-2 (2016–2019) in orbit until recently.

So far, the Apollo programme was the only time in human history when men visited the Moon. In December 1972, the last human set his foot on the Moon and nobody has been there since. The Soviet Union and China have never sent a cosmonaut or taikonaut beyond the Earth orbit.

Apart from this, no human has ever been on another planet, on an asteroid, on a moon other than our own, outside the Earth-Moon system, or anywhere else in the solar system. Once space tourism accelerates, individuals of many countries will have the chance to be the first in space, and many countries might get their first spacefarer up there.

1.3 Space Tourism

The concept of space tourism and the dream of tourist spaceflight has existed for generations. Already back in the 1960s, Barron Hilton proposed hotels in space, and American airline Pan Am had a list of almost 100,000 individuals interested in booking a trip to the Moon. In 1991, however, Pan Am was bought by Delta Airlines after its financial collapse, and never delivered a single person into space. The movie *2001: A Space Odyssey* (1968) presented what flights could look like.

But how many people have been in space as tourists to date? That depends on the definition! So, who has been to space as a self-paying spaceflight participant at an altitude above 100 km as of 2020? The first space tourist in human history was the 60-year-old American engineer and multibillionaire Dennis Tito, who went on a 7 day and 22 h Soyuz mission to the ISS in 2001, apparently paying US\$ 20 million for the trip. He was housed in the Zvezda module of the ISS and was officially only permitted inside the other modules when escorted. The South African Mark Shuttleworth followed Tito in the year 2002 to become the second ever space tourist. The first female self-paying spaceflight participant was American-Iranian multibillionaire Anousheh Ansari in 2006. Between 2001 and 2009, a total of seven paying individuals went to space this way. The others were Gregory Olsen (2005), Charles Simonyi (2007 and 2009), Richard Garriott (2008) and Guy Laliberté (2009). Missions apparently went well and were enjoyable for all of them. All of these trips were organised and sold by the US company Space Adventures, the only business that has successfully brought tourists to space so far. Indeed, many businesses have made millions by selling tickets, but only this one company really managed to keep their promise so far. All seven tourists paid US\$ 20 million or more and were launched in Russian Soyuz spacecraft, as Space Adventures does not have their own rockets or spaceships. In 2009, tourist flights to the ISS were discontinued when the permanent crew size on board the ISS was increased. After the Space Shuttle became retired, the Soyuz was the only spaceship capable of and available for bringing humans to the ISS. All capacities were then needed by the national space agencies, which had priority over tourist flights. This policy lasted until 2019, when Emirati spacefarer Hazza Al Mansouri of the United Arab Emirates Astronaut Programme spent 7 days and 21 h on board the ISS. He became the first Emirati in space and the first Arab on board the ISS. However, again, it is a matter of definition, whether or not he can be counted as a tourist, as he did not cover the costs himself, but was sent by the Prime Minister of his country.

Depending on the definition of space tourism, one might as well count numerous *payload specialists* on board the Space Shuttle as tourists. These are crew members without command or pilot functions, who did not get the same training as the NASA astronauts and were sometimes not even employed by NASA. These individuals, however, did not pay for the flights either, but the costs were often covered by their companies or institutions. Similarly, the Soviet Union, within the Intercosmos programme, brought cosmonauts from allied countries to space—again individuals who did not share the same training and experience as the proper crew. This development might be the future of human spaceflight, as remote control, new technologies and artificial intelligence are expected to reduce the required skillset, and automated flights carrying laypersons become the norm. While astronauts in the Apollo-times were trained for all worst case scenarios and ready to navigate their spaceships entirely by hand and eye, today's systems allow for more flexibility in the crew members' qualifications.

Over the past years, several companies have developed their own spaceships for commercial applications. The Ansari X-Prize of ten million US dollar, announced in 1996 for the first private human spaceflight, was won by the American company Scaled Composites with their *SpaceShipOne* and its carrier airplane *White Knight* on 4 October, 2004. To win the prize, a crewed, reusable spaceship had to reach the Kármán line (100 km altitude) twice within a period of 2 weeks. For a short moment, the success of SpaceShipOne led many people to the assumption that space tourism would now finally progress quickly. However, this was not the case. The prototype of the successor *SpaceShipTwo* named *VSS Enterprise* was destroyed in a crash on 31 October, 2014. Only one of the two pilots survived. A new spaceship, the *VSS Unity*, now owned by Virgin Galactic was presented to the public in February 2016.

In parallel, a number of companies are developing commercial spaceflight capabilities for crewed missions, including SpaceX, Blue Origin and Boeing. As of early 2020, however, only SpaceX has succeeded in bringing a human into space. Also SpaceX's Dragon spaceship has reliably delivered cargo to the ISS for several years now. Boeing's Starliner (previously named CST-100) has reached a very promising state, too, but by early 2020, it was still struggling with technical difficulties in its test flights, and crewed launches were delayed multiple times. It only seems to be a question of time when the first space tourists will depart in a commercial vehicle. As books can never be up-to-date with launch schedules, a QR-code leads to the Spacefacts website, for one of the most accurate and updated global crewed launch schedules available (Fig. 1.3).

Fig. 1.3   The *Spacefacts* website shows which missions and individuals are currently in space, and who will launch and land next. http://spacefacts.de/schedule/e_schedule.htm

Many people like to think that astronauts and cosmonauts fly to space all the time. Crewed flights, however, take place less often than the intense media attention might suggest. Over the last years, usually three to four crewed flights were launched per year world-wide. Three crewed Soyuz spacecraft launched in both 2018 and 2019, respectively. In 2014, 2015 and 2017, four crewed flights took place, while five were recorded in 2012, 2013 and 2016. Activity was much more intense in the 1980s and 1990s. In 1985, for example, eleven crewed missions were launched (nine Space Shuttle and two Soyuz launches). Indeed, there is hope for these numbers to rise considerably in the near future.

1.4 Astronaut Selection

Which criteria need to be fulfilled to be selected as an astronaut? Most of the formal criteria concern education, work history, health, experience and personality of the applicant. But the most important factor is luck! With regard to education and qualification, a great number of people are wonderful candidates. Military, aviation and science backgrounds have in the past promised the greatest chances to be selected. In Germany, eight out of the eleven men who have been to space are physicists. NASA's astronaut corps usually selects the most diverse crowd with all kinds of skills, including engineers, doctors and teachers.

Many people ask themselves if their health is good enough to fly to space. This question can, however, not be answered easily, as there is not simply one clear rule. Certainly, as minimum, at least the requirements for a pilot medical license need to be fulfilled. It is often required as part of the initial astronaut application. So far, only very healthy individuals have been to space, and there is almost no experience with diseases and medical conditions in spaceflight.

In aviation, however, plentiful experience has been gained over many years, that is for now a good approximation. In addition to this, spaceflight exposes participants to weightlessness, hyper-G, radiation, extreme excitement, and a lot more stressors. An applicant usually has no opportunity to argue when anomalies are found during the selection process, and is instead automatically deselected. At the same time, astronauts have flown with minor abnormalities, such as Gilbert's syndrome, a mild liver disorder, where people usually do not have symptoms or limitations.

To choose the very best candidates in the frame of an astronaut selection is called a *select-out*. That means, any candidate with only the slightest flaw is discarded from the crowd of applicants. When it comes to letting experienced spacefarers fly again, to benefit from their experience, insignificant aberrations may be overlooked more easily, as long as they are unlikely to be a risk for mission success. This approach is called *select-in*. The authors believe that many tiny health issues would not be a problem and assume that criteria for tourist flights will become less rigid over the years (see Chap. 5). However, when considering in- or exclusion, many spaceflight-relevant aspects must be taken into account. Selection criteria need to be particularly strict with regard to mental health. Other obvious exclusion criteria are serious heart problems or recent surgery.

It is easily understandable why space agencies select the fittest and best-suitable candidates. This is to minimise risks as much as possible for their expensive programmes paid by tax money. Often, politics and PR aspects play a role, too. The first woman in space was Valentina Tereshkova, a factory worker who was sent to space by the Soviet Union. The political message to the population was that she was not only one of them, but also that normal factory workers could achieve everything, and that everyone is valued.

Which selection criteria for astronaut candidates have played a role in the past? Which steps did the applicants have to complete in the selection process? The European Space Agency last selected astronauts in 2008 (as of early 2020):

- Initially, a formal application had to be submitted including proof of a pilot medical. A university degree, research experience, or when applying as a military pilot, proof of flight experience, and, if available, further evidence of additional skills such as a diving certificate, doctoral degree, Russian language skills, etc. had to be provided.

- Two steps of psychological and professional assessment followed, which also included behavioural and cognitive examination.
- The next level was extensive medical testing by medical specialists of different fields, including blood tests, endurance and cardiac tests, eye and ear examinations.
- A formal job interview by an ESA committee followed. After this stage, only 20 out of more than 8000 applicants were left. The last decision of which individuals to choose from those 20 was mainly political and had a lot to do with their nationality, as the European countries contribute differently to ESA.
- The last step was the announcement of the six new ESA astronauts in training.

NASA's astronaut selection programme has been accepting applications almost on an annual basis since the Space Shuttle days. Pilot and non-pilot individuals may apply. The latest non-pilot application criteria were:

- A bachelor's degree in fields of engineering, biology, physics, computer sciences or mathematics (which ones exactly was not specified).
- At least 3 years of work experience after the bachelor's degree. A master degree and/or PhD were particularly appreciated.
- In case of vision disorders, these needed to be correctable, for example, by wearing glasses.
- Anthropometric dimensions had to fulfil the permitted sizes for spacesuits (these were measured during the interview).
- Applicants had to have US citizenship (having two citizenships was acceptable, too).

Relatively young people were usually selected in the past, but a recent trend towards older spaceflight participants can be observed. One of the reasons why it makes sense to choose more senior individuals for long-term flights to points of interest outside the Earth's magnetic field are the high radiation doses expected there. Exposure to space radiation for a longer period of time significantly increases the likelihood for cancer in later life, that usually does not show up immediately but several years after. When exposing older individuals, their natural remaining lifespan is shorter anyway and the cancer may not even appear anymore, while the life expectancy and quality of life of younger candidates may be affected badly.

Hint for Aspiring Spacefarers It is impossible to predict in detail which spaceflight selection criteria will become most crucial in the future. A lot of luck is needed to be selected anyway! To score well in an astronaut selection, it is not at all required to be an elite athlete or a Nobel Prize winner. The ones chosen are usually good in a great variety of things, and are most of all great team players. Showing no deficits in any one of the tests is most important. Drop-out is often caused by supposedly trivial aspects before consideration of the real strengths of an individual has even taken place. Sometimes space agencies want to find someone with a specific background or in a certain age group, but were unable to communicate this beforehand. To prepare for the selection process, a healthy lifestyle with a balanced exercise routine is recommended, that may include activities such as running, cycling, swimming, going to the gym and team sports. Top results are irrelevant, but overall fitness and consistency are what counts. It is essential to stop smoking and limit alcohol consumption (both habits are reflected in the blood results). Additional skills, such as a pilot or diving certificate, previous expedition experience, such as overwintering in Antarctica or fluency in several languages, help. To prepare for the testing, one might want to repeat and practice school level maths and physics. Basic spaceflight knowledge should be beneficial, too. A good answer to the question why one is the ideal candidate might be helpful in a job interview. To get there, we recommend reading this book!

White men are dominating human spaceflight and have been ever since the beginning. When getting the chance to be decision makers in the future, the readers of this book could choose to optimise their crews' potential as much as possible by selecting the most capable candidates independently of their cultural backgrounds or sexes. Studies have shown that teams consisting of both women and men work best in spaceflight. A diverse team seems to limit aggressions and conflict situations among crew members. In the past, fist fights reportedly took place among male cosmonauts. In several relevant fields (for example, teamwork) women even outperform men. In addition, their body mass is lower and they consume less food and oxygen, saving payload (carrying capacity) and resources.

Back in April 1959, the *Mercury 13* were thirteen female pilots with more than 1000 h of airplane flight experience each, who participated in a privately funded astronaut screening campaign, but never became members of the NASA astronaut corps or flew. Sally Ride much later became the first female NASA astronaut in 1983.

1.5   Checklist Before the Flight

Spaceship built and launch date booked? What is next? Here is a list of issues to consider:

- Cameras ready? Data management planned? See Chap. 2.6 for more.
- Calculated fuel and flight route? Consider reading Chap. 3.3.
- Selected and packed diversified food options in large amounts? Taste changes in space! More details in Chap. 4.3.
- Ordered a spacesuit that is a little oversized? Elongation of the spine by about 5 cm is normal in weightlessness! For details see Chap. 4.4.
- In good shape and exercise routines planned? Check out Chap. 5.4.
- Ready to face the Space Adaptation Syndrome including nausea, vomiting and back pain? Details: Chaps. 5.3 and 5.6.
- Got a message for extraterrestrial life ready? Find inspiration in Chap. 6.8.
- Mission patch ready? (read on)

Before the start of the mission, one item is crucial: the mission emblem. Designing a specific and characteristic patch for each mission has become a wonderful tradition in spaceflight NASA (2012). Apart from the mission name, it may contain the names of all participants. Mission patches are attached to the spacesuits and stuck to all kinds of equipment boxes, devices, notebooks, and the scientists involved in the mission love to have them on their lab or office doors. See Fig. 1.4 for some examples from spaceflight history.

Fig. 1.4 Mission emblems: Apollo 11 (1969, first human on the Moon), STS-71 (1995, first docking of a Space Shuttle to spacestation Mir), Spacelab D1 mission (1985 with German contribution), Soyuz T-9 (1983, fourth crewed flight to spacestation Salyut 7), Shenzhou 9 (2012, beginning of operation of the first Chinese spacestation Tiangong-1) and the FORESAIL satellite patch (with involvement of both authors) (Images: NASA & U. Ganse)

Reference

NASA. (2012). *Human space flight: Mission patch handbook*. NASA. ISBN 978-0981783857.

2

Building Spacecraft

Contents

Let's not waste any time and start: To go to space, some gear will be required. First and foremost, one needs a device that can transport humans into space and keep them alive there: a spaceship!

2.1 Spacecraft Classes

Initially, a clarification of terms: The rockets used in human spaceflight are standing on their launchpad like skyscrapers, yet only a small part at the top of them eventually arrives in space. The lower stages, called the launch vehicle, which are entirely built out of fuel tanks and rocket engines, are required for racing up through the atmosphere and are detached after that. The *proper* spaceship, also called spacecraft, in which spacefarers fly and navigate through

© Springer-Verlag GmbH Germany, part of Springer Nature 2020
B. Ganse, U. Ganse, *The Spacefarer's Handbook*, Springer Praxis Books,
https://doi.org/10.1007/978-3-662-61702-1_2

space, is just the topmost section (see Fig. 2.1). In some cases, it can be further subdivided into the return capsule and the service module.

There is, naturally, more than one type of spaceship: A multitude of different utilisations, special cases and environmental conditions exist, in which spaceships are used. Their shapes and construction styles are therefore equally plentiful. Figure 2.2 shows a rough overview of which classes of spaceships have been built and employed by humans so far. This classification is in no way strict: it would totally be possible to build a spaceplane with the intention to land it on another celestial body, thus making it also a lander. Yet, in spacecraft design it is always necessary to keep weight limitations in the back of the mind, since every kilogram of weight carries massive extra costs. Additionally, every electronic or mechanical system can fail, needs to be maintainable and requires lots of testing on the ground beforehand.

Thus, in order to make spaceship construction in any way economically feasible, the fundamental design principle is always that the simplest, most robust, most streamlined and cheapest method will be used. Hence, in all nuances of spaceship construction this very same principle will occur over and over. It is true for the outer hull of a spaceship, for rocket engines, and even for minute details such as furnishings and interior design.

2.2   A Spacecraft's Hull

Space is cold, devoid of air, abundant in radiation and swirling with small and big meteoroids. In order to survive in this environment, it is absolutely essential to have proper implements to keep unpleasantries outside, while keeping necessary things like air and warmth inside the spacecraft. A spacecraft's hull must fulfil all those functions. The challenge in designing a good spaceship hull is to optimise the requirements, while at the same time being as light and cost-effective as possible. Every kilogram of hull that gets transported to space means 1 kg less of payload mass!

The most common construction material for spacecraft is aluminium, similarly as in airplanes. It benefits from a high level of sophisticated manufacturing techniques available from aviation, it is easily brought into any desired shape by bending, casting and machining and it nicely dissipates thermal stress thanks to its good heat-conducting properties. Recently, however, composite materials containing carbon fibres have found their way into spacecraft construction, for example, in the SpaceShipOne and SpaceShipTwo produced by Scaled Composites and in the Falcon rocket series of SpaceX. The very light and

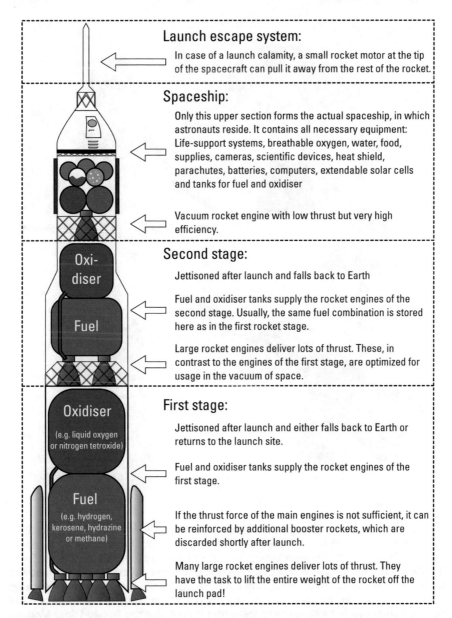

Launch escape system:

In case of a launch calamity, a small rocket motor at the tip of the spacecraft can pull it away from the rest of the rocket.

Spaceship:

Only this upper section forms the actual spaceship, in which astronauts reside. It contains all necessary equipment: Life-support systems, breathable oxygen, water, food, supplies, cameras, scientific devices, heat shield, parachutes, batteries, computers, extendable solar cells and tanks for fuel and oxidiser

Vacuum rocket engine with low thrust but very high efficiency.

Second stage:

Jettisoned after launch and falls back to Earth

Fuel and oxidiser tanks supply the rocket engines of the second stage. Usually, the same fuel combination is stored here as in the first rocket stage.

Large rocket engines deliver lots of thrust. These, in contrast to the engines of the first stage, are optimized for usage in the vacuum of space.

First stage:

Jettisoned after launch and either falls back to Earth or returns to the launch site.

Fuel and oxidiser tanks supply the rocket engines of the first stage.

If the thrust force of the main engines is not sufficient, it can be reinforced by additional booster rockets, which are discarded shortly after launch.

Many large rocket engines deliver lots of thrust. They have the task to lift the entire weight of the rocket off the launch pad!

Fig. 2.1 Schematic diagram of a typical liquid fuelled multi-stage rocket, the most common rocket type to propel crewed spacecraft off the ground. The actual spaceship is situated in the topmost stage of the rocket, everything else returns or plummets back to Earth immediately after launch

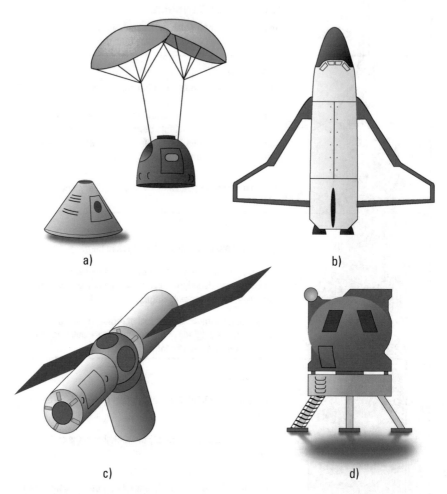

Fig. 2.2 Typical shapes of spacecraft which have been used in human spaceflight: (**a**) Capsules are the most common form of launch and return vessels (they are usually connected to other components until the reentry phase). (**b**) Spaceplanes like the Space Shuttle are used to transport both humans and uncrewed payloads, and can land like an airplane thanks to their aerodynamic profile. (**c**) Bigger spacecraft and structures are constructed from modules in space by docking them together. The modules are launched successively by separate rockets. (**d**) Landers, like the Apollo Lunar Module, are especially designed to touch down on other celestial bodies and launch back to space from them. Their configuration strongly depends on the properties of the target body!

heat-resistant metal beryllium also finds use in spacecraft components, but its toxicity means that it is usually kept away from the crew compartment.

A special case of one-person-spaceship is the spacesuit: Just like a proper spaceship hull, it is designed to protect the human inside from the negative effects of space. It comes with the special requirement that its hull should be built from elastic bendable materials to allow freedom of motion. Oftentimes, layers of Kevlar (as used in bulletproof vests) provide protection against micrometeoroids, Mylar foil (commonly used in emergency blankets) enables a vacuum tight seal and special underwear with sewed-in cooling pipes takes care of thermal regulation (compare Sect. 4.5).

Since spacesuit materials have been optimised and refined over many decades, even spacecraft and spacestation modules are nowadays starting to be assembled from the same flexible materials as opposed to rigid aluminium construction. The inflatable spacestation modules manufactured by the American company Bigelow Aerospace (such as the BEAM module on the International Space Station, see Fig. 6.18) are the most well-known examples of this.

Radiation, Heat and Meteoroids

The Sun is the solar system's biggest source of light, heat and ionising radiation. A total of $1.36 \, \text{kW/m}^2$ of overall irradiance reaches Earth at any given moment, the largest part being infrared, visible and ultraviolet light. Earth's atmosphere (especially the ozone layer) protects us from most of the UV radiation and additionally absorbs parts of the visible and infrared light, so that only about $1 \, \text{kW/m}^2$ eventually reaches the surface. The number changes rapidly when travelling into different parts of the solar system. Flying closer to the Sun, the irradiance per square metre rises as an inverse square law of the distance: In the same way as the apparent size of the solar disc grows when getting closer to the Sun, the energy flow increases. In journeys to the outer part of the solar system, the irradiance decreases likewise. This is the reason why space probes travelling outwards towards Jupiter require significantly larger solar panels than those orbiting Venus.

As opposed to the glaring Sun, the other areas surrounding the spacecraft are cold and dark. Almost no heat reaches the spacecraft from the blackness of space, as its heat radiation corresponds to the thermal glow of a black body at a temperature of −270 °C, only 2.73 K above absolute zero.[1] While the sun-facing side of a spaceship is continuously heated by the Sun, the backside is radiating heat out into cold space. The large difference in heat flux between the Sun and empty space makes it necessary to take heat transport into account when designing spacecraft and potentially add heating and/or cooling mechanisms. Without these, recurrent thermal expansion and contraction may lead to material fatigue and could eventually cause the hull to break and start leaking air.

A hull constructed out of heat-conducting materials could take care of the thermal problems by itself, but this is not always sufficient, especially if the structure is very thin. To actively manage spacecraft temperature, a fluid-based heat transport system can be used (the International Space Station is cooled in this way, using an ammonia-based coolant loop), or heating and cooling devices can be distributed throughout the hull. Another option is to cool via evaporation of volatile substances. All these methods help to counteract thermal stress and minimise material fatigue (thermal management is discussed in detail in Sect. 2.4). The spacecraft of the Apollo programme solved this problem using an even simpler and pragmatic approach: During the uneventful flight period between low Earth orbit and Moon landing, the ship was spinning slowly along its long axis (called the "Barbecue Roll"). Hence, all surfaces were equally exposed to sunlight, levelling out the heating.

Thermal management is also the reason for satellites' typical golden hull colour. As an outer layer, they tend to be wrapped in Mylar foil that has been coated with a thin layer of gold. Gold reflects most of the incoming light, thus keeping the illuminated side cool, while thermally insulating the satellite's cold side.

The design of spacecraft that are intended to spend longer than a couple of days in space further needs to account for the effects of unfiltered solar UV radiation. Certain kinds of plastic, for example, would become brittle and non-UV-resistant colourings would quickly be bleached out. The American flags left on the Moon by the Apollo missions are nowadays completely white, since their 1960s colours were not able to withstand constant exposure to the blistering sunlight!

---

[1]This thermal background radiation is actually the echo of the Big Bang and fills the whole universe equally.

Light (be it visible or invisible) is far from the only kind of radiation a spaceship's hull has to shield off. *Cosmic rays* in particular need to be taken care of! Apart from their harmful effects on humans (see Sect. 5.7), radiation can knock electrons loose from atoms, so that random electric charges appear, and microscopic electronic circuits can be disturbed or damaged. Complex molecules, such as the active compounds in medicine, age at an increased rate under radiation exposure (see Sect. 5.8). On the surface of the Earth, cosmic radiation is warded off almost completely by Earth's magnetic field and atmosphere, but in space its average energy density is around $1.8\,eV$ (approximately $3\,nJ$) per cubic centimetre. While this number lies significantly below the value of visible light from the Sun, fending off this kind of radiation comes with a whole lot of extra problems. As the high-energy charged particles that form cosmic rays easily penetrate deeply into materials (including human tissue), they cannot simply be stopped by thin layers of metal.

When travelling in low Earth orbit, Earth's magnetic field gives reasonably efficient protection against these rays, at least when it comes to less energetic ones. The charged particles follow the magnetic field lines towards the magnetic poles of Earth and plummet into the atmosphere there, causing colourful aurora above the Arctic and Antarctic regions. This deflection effect vanishes when rising into and above the Van Allen radiation belts (between 1000 and 60,000 km above the surface, see Fig. 2.3), as the magnetic field can no longer trap particles at those distances.

Cosmic rays are best protected against, like radioactivity on Earth, by a wall that is as thick, as dense and as massive as possible. Ideal materials to absorb them would be lead or tungsten, but exactly these heavy substances are unpopular for spaceship construction, as mass comes at a premium. For radiation shielding that uses the available mass budget efficiently, it makes sense to employ substances which will anyway be needed in long duration spaceflight. There are, for example, plans to build a spacecraft's fuel and water

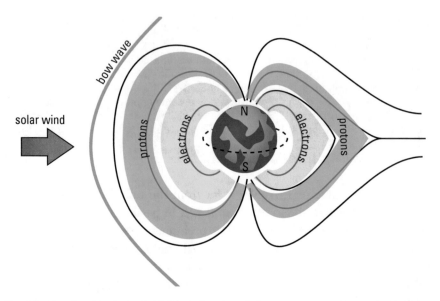

Fig. 2.3 Earth's magnetic field (black lines) and the Van Allen radiation belts (figure is not to scale). The inner radiation belt is populated by electrons (between 1000 and 6000 km above the surface), while the outer belt is filled with protons (between 13,000 and 60,000 km). Human spaceflight in low Earth orbit (dashed ellipse) takes place below the radiation belts and is hence shielded by the magnetic field

tanks as a cylindrical shell around the crew compartment, thus making their liquid contents double as a radiation shield.

Apart from elementary particles, bigger pieces of matter are also buzzing around in space. The sizes of these naturally occurring projectiles range from dust grains that are hardly exceeding the diameter of an aerosol in cigarette smoke, via tennis ball sized rocks all the way to meteoroids comparable to large cities. In near-Earth space, the problem is getting worse due to an increasing amount of human-made space trash: Its size spectrum ranges from burned out rocket stages via screws and bolts, frozen drops of rocket fuel to specks of paint. The danger lies in the immense speeds at which all of them travel around the Earth. Speeds in a low Earth orbit are around 7 km/s, while our planet itself moves around the Sun at a pace of 30 km/s. Encountering an object in a similar but oppositely directed orbit means the collision speed is even higher. Objects from outside the solar system are typically even faster than this. These velocities are many times higher than the speed at which armour-piercing ammunition is being fired (approximately 1 km/s)! It is easy to imagine that multiple metres of steel armour would be required to resist the impact of a meteoroid. Transporting a mass like this is completely unrealistic. Instead,

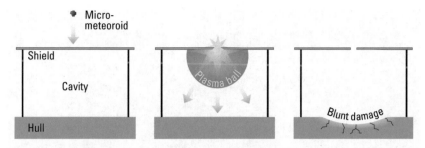

Fig. 2.4 Sketch of the mechanism of a Whipple shield against micrometeoroid impacts. A thin foil suspended away from the hull causes impacting particles with high velocity to disperse into clouds of plasma. This cloud spreads as it traverses the cavity and creates blunt damage or deformation over a wider area of the hull, instead of piercing through it

the current strategy is to track larger objects with ground-based radar systems and actively avoid them. But it will not be possible to track and evade every marble ball sized projectile moving at or above 30 km/s.

For meteoroids just a little bit smaller than this, with a diameter of a centimetre or below, a very effective and at the same time lightweight defensive structure does exist: The so-called *Whipple shield*, shown in Fig. 2.4, consists of a thin foil (typically aluminium or a ceramic composite structure), which is suspended a couple of centimetres outside the spacecraft's hull. When a meteoroid impacts this foil at a speed between 3 and 18 km/s, it punches right through it. In that instant however, its enormous kinetic energy gets converted to heat and the particle shatters into a cloud of plasma. By the time this plasma cloud has traversed the cavity towards the next hull layer, it has sufficiently expanded. The impact force is spread over a larger area and no longer causes significant damage. To remain effective for even larger and faster particles, the same setup can be repeated in multiple layers, and tear resistant materials (such as Kevlar) can be added.

The problem of impacts by intermediate-sized projectiles, between the size of a marble and a tennis ball, remains. If hit by such a projectile, one can only hope for a clean punch through the spacecraft, causing no damage to any vital systems, and that the holes left are quickly patched up. Figure 2.5 shows an example of such an impact, which caused a sizeable hole in a radiator panel of the Space Shuttle Endeavour. Luckily, there was no further damage. Modern spaceship hulls include self-sealing materials, similar to those in puncture-resistant bicycle tires or self-sealing fuel tanks, to maintain airtightness in such cases.

Fig. 2.5 Impact crater of a micrometeoroid or space trash particle. This projectile punched through a radiator panel of the Space Shuttle Endeavour. The impacting particle was probably about pea-sized, resulting in a half-centimetre entry hole (Image: NASA)

Atmospheric Reentry Heat Shields

When reentering into the atmosphere, a spacecraft's hull is faced with quite different challenges than in space. In the vacuum of space, very low amounts of force are acting on the hull material. Furthermore, thermal problems are not so much that the spacecraft is heating up a lot, but that heat is hard to get rid of. An atmospheric reentry, however, includes a rapid transition into a hot and dense environment, and turbulent airflows cause external forces and strong accelerations. At the end of the reentry process, the spacecraft finds itself under "normal conditions", as they are known on Earth, and will have to be able to deal with them, too.

The enormous amount of heat created in an atmospheric reentry poses the biggest technological challenge. Where does this heat come from? At launch, a huge amount of rocket propellant was used to give the spacecraft the kinetic energy it needs for its travel to orbit (see Sect. 3.3), and this energy has to be disposed of for a safe landing back on our home planet. In principle, the slowdown could also be performed using rocket engines, but these would have to be similar in size and power to the ones used at launch! A much more economical way to remove kinetic energy is to use the friction of the atmosphere. Since the spacecraft is coming in with an orbital speed that is multiple times the speed of sound, there is plenty of friction created (from the point of view of the reentering vessel, the situation looks like it encounters a head wind hitting it faster than the speed of sound). Friction means that air molecules are rubbing over the hull's surface and heat it up, thereby converting kinetic energy into heat. One could envision an aerodynamically shaped spaceplane with wings and flight control surfaces, such as flaps at the back of the wings, to be the optimum profile for this. Slowly descending from the thin outer layers of the atmosphere into progressively denser ones (compare Fig. 5.23), the glide angle could be adjusted to keep friction, and thus heating, under control. Unfortunately, due to the immense amount of heat, this descent would take a very long time. Luckily, there is a better approach!

When the first wind tunnel experiments for reentry vehicle design were conducted in the 1950s, engineers found to their astonishment that perfectly aerodynamic test bodies were actually heating up much more than blunt-shaped items they placed into their supersonic (faster than the speed of sound) air currents. In other words: objects with high wind resistance were heating up less than objects with little surface friction. The reason behind this paradox was found to stem from shock waves that are formed when a supersonic air flow encounters an obstacle. In these shock waves, the flow suddenly drops from supersonic to subsonic velocity, and as a result, air gets compressed and heated. The heating can be so immense that oxygen molecules get ripped apart, and the air turns into plasma, which radiates heat through infrared light. In an aerodynamic configuration, which optimises smooth and laminar air flow over its surfaces, these shock waves are in direct contact with the surface material (compare the arrow-shaped outline in Fig. 2.6a). The result is strong heat transfer from the shocked air onto the hull, in addition to the direct friction heat.

In blunt objects, such as the other models shown in Fig. 2.6b–d, the air flow accumulates and stagnates in front of the object, so the shock wave assumes the form of a bow wave. The shock wave is no longer in direct contact with the reentering body, but has a standoff distance determined by characteristics of the blunt body, such as curvature and tapering. Incoming air from the supersonic upstream direction, after having been heated by the shock, further decelerates and moves around the body at a slow speed, if it even gets close to it. The vast majority of air gets deflected to the sides and never has the chance to transfer its heat to the hull. Both friction heat and conducted heat amounts are therefore minimised, and it is only the infrared heat radiation of the shock wave, which still transfers significant amounts of heat to the blunt side of the reentry vehicle.

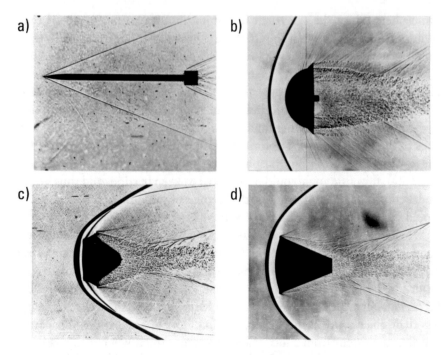

Fig. 2.6 Wind tunnel pictures showing the airflows around different shapes of reentry bodies. (a) In an aerodynamically shaped body, shock waves of supersonic air flow directly contact the spacecraft, and a large amount of friction heating occurs. (b)–(d) Using more blunt form, the shock waves gain a larger standoff distance from the reentry vehicle, and direct contact with the heated air is significantly reduced. These pictures show the development of spacecapsule designs in the Mercury programme between 1953 (a, b) and 1957 (d), as well as a missile nose cone (c) (Image: NASA)

The first manned spacecraft, Vostok 1, with Yuri Gagarin as the first human to reach space had a spherical reentry body. This design was considered advantageous, as the rotation of the sphere meant that different sides of the hull were being heated, and an equal temperature distribution was achieved. Exactly this tumbling and rotation, however, is extremely uncomfortable for the spacefarers inside the capsule, as they are subject to unpredictable acceleration forces (rapidly changing G-forces). Newer reentry capsules are typically constructed in a sphere-cone structure, in which a spherical blunt side connects to a cone segment. This shape automatically orients itself into the reentry airflow, so that acceleration is always acting in the same direction. Forming the shape slightly asymmetrical also allows for a limited degree of reentry flight control, by turning along the long axis of the craft.

The Space Shuttles (and likewise, the Russian Buran spaceplane) had an unusual shape of reentry body: Their complete, airplane-shaped orbiter was reentering into the atmosphere. This seems to contradict the earlier observation that aerodynamic shapes are bad for heat management in the atmosphere. To still get the benefits of a blunt reentry shape, these spacecraft were *not* entering the atmosphere nose-first, but were positioned so that their flat underside was facing forward. In order to maintain this unstable alignment, the onboard computers had to perform precise steering commands to the control surfaces and thrusters. During the last flight of the Space Shuttle Columbia, a damaged heat shield tile on the starboard wing edge caused the control surfaces on that side to fail. For this reason, the shuttle went into an uncontrolled tumbling motion, where it no longer managed to keep the shock wave away from the hull—leading to the destruction of the spacecraft and loss of the entire crew. As a result, NASA decided to return to "conventional" spacecapsule designs, even though this decision means sacrificing the controlled landing ability of the spaceplane shape.

In addition to the geometry of the spacecraft, also the choice of heat shield material has a big effect on its reentry properties. Two main technologies are used (compare Fig. 2.7):

- *Ablative heat shields* are made from materials, which decompose through pyrolysis in the reentry heat (Fig. 2.7a). The gases evaporating from their surface are pushing against the incoming airflow, further increasing the standoff distance of the incoming shock wave. In addition, the produced gases are close to the boiling temperature of the ablator material, and thus significantly colder than the surrounding blazing air. The Mercury programme was the first to employ this kind of shield, made out of fibreglass and epoxy resin.

  Modern heat shields are made out of a mixture of carbon and phenolic resins, which are either applied as a powder mixture, or, in the case of the PICA-X heat shield of the SpaceX Dragon capsule, as carbon fibre layers soaked in resin. As the resin decomposes, carbon particles are transported out of the heat shield, reacting with the incoming air. As opposed to glass fibre particles, that simply melt and evaporate, carbon particles brightly glow as they burn away, similar to a white glowing flame in a hot fire. This carbon flame is opaque to infrared radiation, meaning that the heat radiation of the shock wave is reflected away and no longer reaches the spacecraft. The overall heat transfer is hence reduced even further. While ablative heat shield materials are light and relatively cheap, they come with

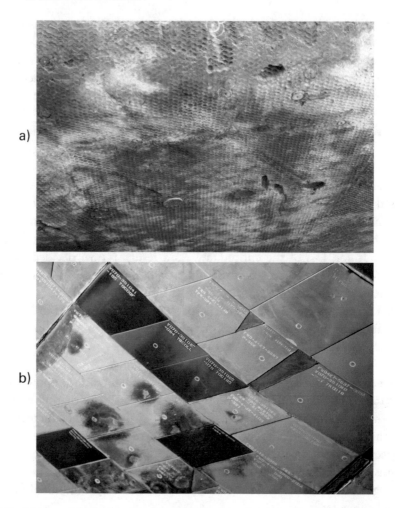

Fig. 2.7 (a) Heat shield of Apollo 12 after return to Earth (Image: NASA) and (b) tiles of the Space Shuttle Atlantis heat shield system (Images: B. Ganse)

the disadvantage of being single-use products and have to be replaced after every flight.

- *Thermal soaking heat shields* take a very different approach: They do not dissipate the heat pulse of reentry by quickly transporting it away from the spacecraft, but store most of the heat in special materials and slowly radiate it away afterwards. This requires a very high heat capacity, but at the same time very low heat conductance into the inside of the spacecraft.

The first Mercury capsules attempted this with a shield made of solid beryllium (which was replaced with the aforementioned ablative shield before the actual crewed flights), whereas the Space Shuttle heat shield was made up of tiles of a special ceramic material (Fig. 2.7b). The Space Shuttle ceramic was able to absorb a major fraction of the heat of an atmospheric reentry, equalling 3000 GJ of thermal energy (this could be used to heat a home over 40 years). After landing, its outer skin had a temperature of over a thousand degrees Celsius, while the inside of the heat shield was no warmer than human skin! At the same time, it managed to be less dense than Styrofoam. After every flight, multiple weeks of waiting were required until the tiles had completely cooled to ambient temperatures. The less heated upper side of the Space Shuttle orbiter was covered in Nomex tissue, the same material that protective clothing for firefighters is made of (Fig. 2.8). These systems are reusable. Ideally, once the stored heat has dissipated, they can be launched right back into space, but being overall much more expensive, typically heavier and proving to be quite damage sensitive, ablative heat shields are nowadays the dominant technology.

Heat shield efficiency can further be augmented through active cooling, e.g. by spraying water streams from nozzles directly into the shock waves.

Fig. 2.8 Thermal protection on the side of the Space Shuttle Atlantis. More strongly heated areas are equipped with heat shield tiles, less strongly heated ones are padded with Nomex tissue (Image: B. Ganse)

Unfortunately, all research results about this technology appear to be secret! One can hence assume that this technique is employed in the reentry bodies of ballistic missiles.

Another field of heat shield research deals with "hot structures". The idea behind them is to design a much larger part of the spaceship, including the entire hull structure, to heat up significantly during reentry. Heat insulation is then mounted on the inside of the hull. While this requires heavier materials (such as steel in place of aluminium), the reduced weight of heat shielding can bring forth an overall weight loss. SpaceX's Starship, constructed entirely out of stainless steel, is the first test of a hot structure design for crewed spaceflight.

The most extreme atmospheric reentry was performed by the Galileo space probe, when it entered into the atmosphere of the gas planet Jupiter in July 1995. It reached the top of the atmosphere with a staggering speed of 47 km/s and slowed down to subsonic speeds within 2 min. In this process, 80 kg of ablative heat shield material evaporated! Immediately after, the probe further slowed down with parachutes, and descended into Jupiter, transmitting valuable scientific data from the lower strata of the atmosphere.

A spacecraft also goes supersonic when ascending through the atmosphere after launch, which causes the same kind of shock wave to form in front of it. In comparison to atmospheric reentry, however, this shock wave has much less extreme properties. While the velocity increases beyond the speed of sound, atmospheric density rapidly drops when reaching higher altitudes, so no advanced heat shield materials are needed. If the payload contains delicate structures (such as antennas or solar panels), it can still be a good idea to encase them or the entire craft in a protective shell, called the *payload fairing*, to prevent any damage. These can be simple plastic or metal shells.

Landing

Once a spacecraft has gotten rid of most of the kinetic energy it had on orbit and survived the heat and turmoil of atmospheric reentry, the next phase of the flight begins at an altitude of a few kilometres, while falling like a rock. Now, technical solutions are required to make the rest of the journey to the surface as comfortable as possible and most importantly prevent the vessel from forming an impact crater.

Spaceplanes like the Space Shuttle, the Soviet Buran, the Air Force's secretive X-37B or the privately developed SpaceShipOne are now having their hour to shine. Using the full advantage of their aerodynamic form, they land like normal aircraft and thus make up for their more complicated reentry performance. All they need is a landing strip, or sufficiently flat piece of land (the Space Shuttle's first landings took place on the salt pans of Edwards Air Force Base).

For capsules and other reentry bodies without an aerodynamic shape, parachutes are the most common piece of technology to slow down before landing. They pack compactly to save space, and are dragged out from their enclosure by a smaller drogue chute. Since their material strength is limited and opening behaviour is strongly dependent on airflow, care must be taken to open the parachutes at the correct time. If opened too early, they might tear off because of the speed. If opened too late, there might not be enough time for the parachutes to properly unfold. The field of parachute engineering is full of rules of thumb gained by experience. Plenty of wind tunnel and drop tests are required before this method is trustworthy enough for humans to use. Even then, multiple redundant parachutes are usually deployed to assure a safe landing in case of a single canopy's failure.

A paraglider offers an interesting compromise between aerodynamic landing and a blunt body. Just like a parachute, it unfolds from a compact packing in free fall but allows for active control of the glide direction. In the 1960s Gemini programme, a paraglider landing system was originally considered but reliability concerns made a parachute appear to be the better choice. The Russian Kliper design (envisioned as a small spaceplane with a lifting body form instead of wings) likewise included a paraglider landing concept, but never got off the drawing board.

In both forms of textile assisted landings, be it parachute or paraglider, it is a good idea to dump all dangerous or explosive goods (like remaining fuel or noxious coolants) overboard. Any other additional weight, like a single-use heat shield, can also be dropped to facilitate the landing.

All aforementioned approaches to landing are based on the assumption of a dense atmosphere, like the one on Earth, on Venus or on Saturn's moon Titan. When trying to land on celestial bodies with a tenuous atmosphere or on those that are not surrounded by a gaseous shell at all, landing velocity has to be shed using other means. The most obvious way to do so is to use rocket engines, in what is called a powered or propulsive landing. The lunar landings of the Apollo programme have been the only time humans landed using rockets exclusively, whereas most uncrewed space probes have touched down on their destination bodies using this method. The Russian Soyuz capsules, already slowed down with parachutes, fire small solid rocket boosters during the last three metres of the approach, in order to further cushion the contact onto the hard ground of the Kazakh steppe. The crewed capsule concept for SpaceX's Dragon spacecraft allows for a fully powered landing option, including parachutes only as a fallback mechanism.

The landing process is already a challenge, if the lander body has a compact shape (like a capsule). However, it gets an order of magnitude more precarious when landing with an elongated body, such as an entire rocket stage. Until very recently, this was considered an impossible feat, as the elongation makes it dynamically unstable—every bit of tilt in any direction causes it to tip over rapidly. Unpredictable wind gusts in Earth's atmosphere further destabilise the object. Thankfully, nonlinear control systems have made remarkable advances in recent years (thanks partly to the development of quadrocopter drones, which likewise fly in an unstable configuration, balancing on their own downward thrust). Landing entire rocket stages has become a regular trick of the trade that started with SpaceX's Falcon rocket family.

Unless the landing is targeted towards a mass of water, snow or other soft medium, suitably designed landing legs are a must. They should be able to absorb all remaining downwards momentum at the end of flight, while at the same time preventing the spacecraft from bouncing back into space.

Within the landing legs, the so-called *crush cores* are typically performing both these functions. They are segments built from a lightweight material, such as aluminium sheets that are constructed into hexagonal thin-walled cells, which plastically deform and collapse under stress (Fig. 2.9). The Apollo lunar modules successfully used crush cores on the Moon, and Falcon rockets' landing legs survive their sometimes rough touchdowns thanks to them. The notable disadvantage is the fact that they are single-use: once compressed, the crushed material cannot be expanded again. Replacing them after every flight or leaving them behind on a distant celestial body may not always be practical nor sustainable.

A reusable shock absorber can be built as a pneumatic system with a compressed gas cylinder, similar to the ones used in cars. Getting these to work reliably in vacuum, however, turns out to be a lot more complicated, as more elaborate sealants are needed. The overall shape is thus less compact than inside an atmosphere. Another reusable design could be based on electromagnetic eddy current brakes similar to the ones used in trains and trucks. These have, as yet, been too bulky compared to single-use crush cores.

When landing on a very small body like a comet or asteroid, one might encounter another problem: Since gravity on the surface is very weak, the spacecraft needs to be anchored down to keep it from bumping off and drifting back into space. This is especially true for bodies from which volatile gases escape into space. There is a vast abundance of ideas how to perform this anchoring, from shooting a harpoon downwards to drilling screws into the

Fig. 2.9 Example of a crush core material, which can absorb kinetic energy by plastically deforming. This specimen was used in the prototype of an uncrewed Surveyor Moon lander (Image: NASA)

surface, but none have so far proven to be overly successful. The European Space Agency's Rosetta mission to the Comet 67P/Churyumov–Gerasimenko tested a number of different methods with its lander Philae without any success in preventing it from bouncing.

A number of space probes and rovers landing on Mars have used yet another approach: airbags. By deliberately bouncing off the surface multiple times, they got rid of their excess momentum. However, for crewed spacecraft, this option is not applicable, as the airbag shell material would have to be incredibly tearproof to prevent bursting. Uncontrolled bouncing off the surface would also shake the crew around, as every rebound causes an unpredictable jerk in a random direction. This would likely lead to injuries and loss of consciousness.

In the end, all these considerations are necessary only if the spaceship is intended for landing. Spacestations, communication satellites, probes and rocket interstages can reduce mass significantly by ignoring aerodynamics, wind- and weatherproofness (more about weather in Sect. 6.4), heat shields and landing equipment.

Typical Shapes and Constructions

Not only spacecraft intended for reentry have a shape that is governed by the laws of physics. Certain practical construction schemes have likewise been established for vessels that reside in space permanently (see Fig. 2.2).

Every rocket or other launch mechanism that flings spacecraft up from the surface of the Earth has some restrictions on the maximum size of its payload. For rockets, the size of the payload fairing around the tip determines this limit. Also spacestations assembled in orbit are constrained in their design, as every individual element still needs to fit within the typically cylindrical maximum volume of the launcher. This leads to spacestations' characteristic barrel-shaped module structure.

The Russian (Salyut, Mir) and the Chinese (Tiangong) as well as the International Space Station consist of modules plugged together, and contain an inhabitable volume for crew inside of them. Depending on their function, the modules are crammed with laboratory, life support or propulsion systems.

The Russian TKS spacecraft (*Transportnyi Korabl' Snabzheniia*, transporter for supply purposes) surprisingly often provides the basic shape for station modules. It was originally intended as a replacement for the ageing Soyuz spacecraft and supposed to become the workhorse of the Russian spaceflight programme. As such, its design carries all the marks of a fully equipped spaceship, including reentry capsule, propulsion, life support and power supply systems (Fig. 2.10). However, it was never actually used as a crewed spaceship. After some initial test flights as a cargo transporter for the Salyut and Almaz spacestations, the decision was made to simply leave one docked permanently

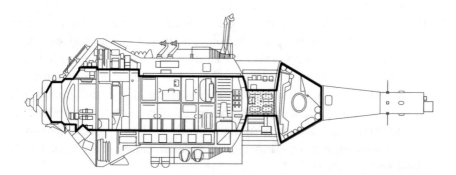

Fig. 2.10 Cross section through a TKS spacecraft, which serves as the basic design for a large number of spacestation modules. Its original concept shown here contains a detachable reentry capsule, located on the right (Image: NASA)

to Salyut 7. It increasingly found use as the basic building block for modular spacestations. In the Mir station, the Spektr, Kvant-2 and Priroda modules were based on a TKS design, while in the International Space Station, the central Zvezda module and the (still unlaunched) Nauka module contain its heritage. The concept of the possible future Mir-2 station by the Russian State Space Corporation Roscosmos again contains a TKS-based central service module.

Future long distance missions to other planets and permanent space habitats will most likely use similar module-constructed spacecraft, probably with an important difference. For very extended stays in space, these modules will potentially not be assembled in a linear manner, but form an overall circular shape or at least large segments of a circle. Artificial gravity is created by spinning the entire structure around the centre. While science-fiction depictions are full of this style of spaceship, not a single such craft has been built. The real-life practicability of such an approach is therefore largely unexplored.

From a technological perspective, a rotating spacecraft would nowadays be fully feasible and helpful in counteracting the adverse effects of prolonged stays in weightlessness (see Sect. 5.9). In order to get anywhere close to simulating Earth's surface gravity, however, the spacecraft needs to be considerably larger than any spacestation built so far or spin at a substantial rotational velocity. Experiments have shown that more than two revolutions per minute are causing discomfort in long duration stays, and spacecraft docking processes get significantly more difficult the faster the docking target is spinning. In-between designs, in which a part of the spacecraft is stationary and other (ring-shaped) parts rotate are certainly possible, but require a reliable, airtight coupling element between the stationary and the moving part. While all of these are solvable problems, they have so far prevented the use of spin gravity in human spaceflight.

2.3 Engines, Thrusters and Rockets

If there is one adjective that properly describes space, it is the word *empty*. Even the best laboratory produced vacuum contains a lot more atoms per cubic centimetre than any point in space. This makes the vast majority of propulsion methods of daily life on Earth infeasible. One can neither row, nor roll, nor be pushed or pulled by a propeller. There simply is nothing to interact with and cause movement, meaning there is no mass around to exert a force on— with the exception of the mass that the spacecraft carries by itself. The only

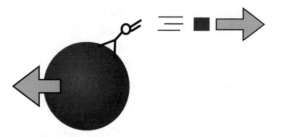

Fig. 2.11 The principle of recoil: Throwing a mass with high velocity into one direction pushes the thrower into the opposite direction. The product of mass and velocity, called the momentum, is equally large on both sides. For this reason, rocket engines try to eject mass with as high a velocity as possible

straightforward possibility to start moving into a desired direction in space is hence to throw out some of the onboard mass into the opposite direction!

The physical law behind this is called momentum conservation. If a body is initially sitting at rest in a vacuum (momentum = 0), and an amount of mass gets thrown out into one direction, the body has to react by moving into the opposite direction to maintain an overall zero momentum (Fig. 2.11). This also implies: The higher the velocity at which mass gets thrown out of a spacecraft, the faster it accelerates.

Actually flying a spaceship by tossing out heavy rocks would obviously be of little practical use. Instead, rocket engines are burning propellant and ejecting hot combustion gases with immense velocity. It was only after the invention of rockets that humans were able to travel into space.

Crunching the Numbers: The Rocket Equation

The question which celestial objects are reachable with a spacecraft is equivalent to the question how much speed a spacecraft can generate using its engines to overcome the gravitational pull of a planet. The answer to this question is given by the *rocket equation*, which was formulated already in the year 1897 by the Russian elementary school teacher Konstantin Tsiolkovsky:

$$\Delta v = v_e \ln \frac{m_{\text{initial}}}{m_{\text{end}}}$$

This equation calculates a quantity called Δv ("Delta v"), describing how much a rocket can change its velocity by ejecting part of its mass (the rocket fuel exhaust). Simply drifting around in space at some fixed velocity consumes no fuel at all, but accelerating, decelerating or steering demand a change in velocity; Δv

(continued)

describes the overall rocket manoeuvring budget. A rocket that can change its velocity by 4000 m/s can do more manoeuvres and reach more destinations than a rocket with Δv of 2000 m/s.

Both the initial mass $m_{initial}$ and the end mass m_{end} of the spacecraft appear in the equation along with the exhaust speed v_e. This means: Taking more mass along in a spacecraft also requires more mass to be ejected in order to reach the same velocity change. This is the reason why rockets quickly get very large when they are supposed to transport humans to other celestial bodies. Every kilogram of additional payload demands multiple kilograms of extra propellant that again needs to be lifted using more propellant. The amount of thrust the rocket produces does not appear in the equation at all! Only the speed, at which the rocket ejects the combustion products, determines the overall velocity change capabilities (Δv). Hence small but highly efficient engines are preferable once a spacecraft has left Earth's atmosphere.

A further consequence of the equation was the invention of the multi-stage rocket. While, in principle, one big single-stage rocket could carry the fuel required for the entire flight in big tanks, the resulting large mass of their metal structures would cause $m_{initial}$ and m_{end} to permanently lie close together. This does not allow the resulting velocity change to be very large at all! By ejecting emptied rocket stages, the mass is consecutively reduced, and m_{end} ends up being only the payload mass for the final stage.

For example, the rocket equation can be used to calculate how well the Saturn V rocket would have worked, had it been using a single-stage design (but propellant amounts and overall mass would have been the same as compared to the multi-stage version). The entire vehicle weighed a massive 2970 tonnes at lift-off, burning kerosene and oxygen to create an exhaust at a speed of 2.58 km/s. Assuming all propellants of the rocket had been burned away, and only the structural mass of 261 tonnes remained, the rocket equation gives

$$\Delta v_{single} = 2.58\,\text{km/s} \times \ln\left(\frac{2970\,\text{tonnes}}{261\,\text{tonnes}}\right) = 6.1\,\text{km/s}.$$

The actual velocity gain of a multi-stage Saturn V can be calculated as the sum of the individual stages: At the end of the first stage burn, the rest of the rocket including the upper stages still weighed 810 tonnes:

$$\Delta v_1 = 2.58\,\text{km/s} \times \ln\left(\frac{2970\,\text{tonnes}}{810\,\text{tonnes}}\right).$$

The empty first stage, still weighing 130 tonnes, was jettisoned, and the remaining 680 tonnes of rocket were lifted by the engines of the second stage, which had a much higher exhaust velocity of 4.13 km/s. The end mass of this stage was 224 tonnes:

$$\Delta v_2 = 4.13\,\text{km/s} \times \ln\left(\frac{680\,\text{tonnes}}{224\,\text{tonnes}}\right).$$

Finally, also the empty second stage was ejected, getting rid of another 40 tonnes and reducing the third stage mass to 184 tonnes. At the end of its burn, only 91

(continued)

tonnes of spacecraft and lunar lander payload remained:

$$\Delta v_3 = 4.13 \, \text{km/s} \times \ln\left(\frac{184 \, \text{tonnes}}{91 \, \text{tonnes}}\right).$$

Altogether, the three stages' Δv values summed up are

$$\Delta v_1 + \Delta v_2 + \Delta v_3 = 10.2 \, \text{km/s}.$$

Reaching a low Earth orbit requires a velocity change of about $\Delta v = 9.5 \, \text{km/s}$ (in order to overcome gravity and counter atmospheric friction), so the single-stage version would not have been able to fulfil the job. The actual Saturn V rocket still had some propulsive capacity left over, which sent the payload on its way towards the Moon.

The rocket equation, however, is only valid—as its name implies—for rockets. Solar sails, space elevators and other more exotic methods of propulsion can circumvent these limitations.

Big Rocket Engines for Launch

Travelling into space means soaring up to a height above the atmosphere, at least 100 km in altitude. Many different ways lead to space: Lifting spacecraft up with hot air balloons, carrying them on jet-powered aircraft and even raising them to the top of the atmosphere with ridiculously tall elevators are among the possibilities. Rocket engines, however, are the only known technology that can continuously and reliably lift heavy loads all the way from sea level into the vacuum of space. The usual approach is to exclusively rely on rocket engines to get into space, instead of combining multiple different technologies in the launcher (which is more complicated and failure prone).

Notable exceptions to this rule are the rockets of the Pegasus family produced by Northrop Grumman Innovation Systems, which are first lifted to an altitude of about 12 km by a cargo plane before igniting their rocket engines. The SpaceShipOne and SpaceShipTwo of Virgin Galactic are likewise launched with a specialised carrier plane.

Every rocket engine consists of a similar set of building blocks. It contains a combustion chamber, where propellant gets converted into hot gas through a fiery chemical reaction and a nozzle, through which these hot gases are ejected in order to create thrust. Shapes, amounts and details of these elements, however, can vastly differ, depending on the kind of fuel used in the rocket (compare Fig. 2.12):

Solid fuel rocket

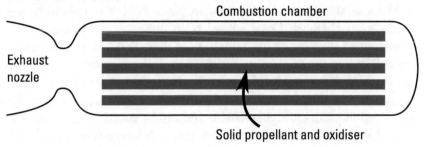

Combustion chamber

Exhaust
nozzle

Solid propellant and oxidiser

Liquid fuelled rocket

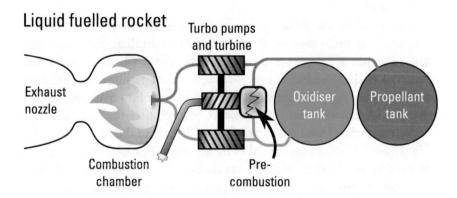

Turbo pumps
and turbine

Exhaust
nozzle

Oxidiser
tank

Propellant
tank

Combustion
chamber

Pre-
combustion

Hybrid rocket

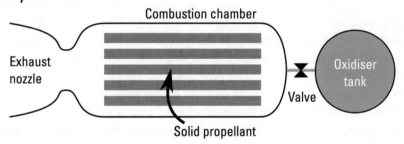

Combustion chamber

Exhaust
nozzle

Oxidiser
tank

Valve

Solid propellant

Fig. 2.12 The three dominant rocket engine technologies: solid rockets, liquid fuelled rockets and hybrids. In a solid rocket, the mixture of fuel and oxidiser is simply ignited and burns away inside the combustion chamber, ejecting its hot exhaust gases through the nozzle. In liquid fuelled engines, turbo pumps, which are typically driven by the exhaust gases from a precombustion chamber, press liquid fuel and oxidiser into the combustion chamber from separate tanks. A hybrid engine contains fuel in solid form, whereas oxidiser comes in liquid or gaseous form, and the speed of the combustion process can be regulated through a valve between the two

- *Solid rocket motors* use propellants which are solid under normal conditions. A block of this fuel is firmly fixed in place inside the (relatively large) combustion chamber. Once ignited from one end, the propellant mass slowly burns away over a fixed duration of time. Small solid rocket motors are well known from fireworks, and the same technology can be scaled up to the size of the towering Solid Rocket Boosters of the Space Shuttle. Thanks to their uncomplicated structure, they often are a convenient solution: They are cheap to produce, highly reliable and can be stored for long periods of time. One significant shortcoming of this technology is that solid rocket motors, once ignited, cannot be shut down until they have completely burned out!

  Compared to the other types, solid rocket motors burn much more irregularly, as their large combustion chambers oscillate like organ pipes when the hot gases stream through them. These vibrations make their thrust more shaky and turbulent. A crewed launcher using only solid rocket engines makes for a very uncomfortable way to get into space, to the point that NASA cancelled its Ares 1 rocket project after the first test flight, because astronauts would not have been able to read the cockpit displays.

- *Liquid fuelled rockets*, on the other hand, employ propellants in liquid or gaseous form, which are resting in tanks outside the combustion chamber. In order to produce thrust, these propellants need to be continuously pumped into the chamber, where they ignite and again create hot exhaust gases that drive the rocket. Propellant flow and thus rocket thrust can be throttled and stopped at any point and in some designs, re-starting the engine in flight is a possibility. These bonuses are offset by a considerably higher technological complexity of the system. The combustion chambers of a typical rocket as it launches from the surface of the Earth contain pressures far above 100 bars. The propellant mixture needs to be pumped into the combustion chamber at an even higher pressure than this, and it needs a velocity higher than the speed of the flame front, so that the flame does not reach the tanks. Therefore, complex turbo pumps and extremely robust plumbing are mandatory.

  A turbo pump works in a similar way as any other compressor (like that of a jet engine or the turbocharger in a car), where turbine blades attached to spinning wheels shovel a liquid downstream with ever-increasing pressure. A considerable source of power is needed to operate the pump! The most common setup to supply the turbo pumps with power in a rocket engine is called the gas generator cycle. In a gas generator, fuel and oxidiser burn in a smaller precombustion chamber and their exhaust drives a turbine wheel, which is connected to two turbo pumps via an axle. The turbo pumps then compress and accelerate fuel and oxidiser into the combustion chamber (for a real-life version of a rocket engine with a gas generator cycle, see Fig. 2.14). The turbo pumps can alternatively be electrically driven, or by using an additional compressed gas, such as helium.

Despite their complexity, liquid fuelled engines are clearly superior to solid rocket motors in terms of efficiency and total thrust. These are therefore the dominant rocket engine technology used in human spaceflight so far.

A huge selection of possible propellant combinations are available for liquid fuelled engines, each having pros and cons. Liquid oxygen and hydrogen (LOX/LH2) were powering the main engines of the Space Shuttle and the Delta rocket family. They have a high combustion temperature and excellent fuel efficiency, but come with the disadvantage that liquid hydrogen is not very dense and thus requires bulky tanks. Prolonged contact with liquid hydrogen also makes many metals brittle, leading to more complicated metallurgy of the engines.

A combination of liquid oxygen and kerosene (LOX/RP-1), as used by the Soyuz and Falcon rockets, has a much higher density and allows rockets to be built much slimmer, at the expense of some efficiency. Both these fuel mixtures contain liquefied gases that are cooled to very low temperatures and quickly evaporate in contact with air, making their storage and plumbing quite complex.

The Chinese Long March and Russian Proton rockets, on the other hand, rely on a mixture of nitrogen tetroxide and dimethylhydrazine (N_2O_4/UDMH). These propellants are much easier to handle as the components are liquid and stable at room temperature. Both are unfortunately extremely toxic!

The book *Ignition!* by Clark (1972) gives a very thorough and entertaining treatise on the history of liquid rocket propellants and their properties. It is strongly recommended for everybody interested in knowing more about rocket fuel mixtures.

- *Hybrid rockets* form a halfway point between solid and liquid fuelled engines. In this setup, one part of the propellant resides inside the combustion chamber in solid form, whereas the second part is liquid or gaseous and gets injected from a separate tank. This combines some of the conveniences of solid and liquid fuelled rockets: Their technological complexity lies only slightly above solid fuelled rockets and, in the case of a gaseous propellant, requires neither turbo pumps nor other failure-prone moving parts. These engines can be cut off at any time, and throttling them is possible to a limited extent, beyond which the low flow rate of the liquid component causes the engine to flameout.

Nonetheless, even hybrid rockets have their disadvantages: The fuel mixtures available for this form of construction have significantly lower specific energy and are thus far less efficient than liquid fuelled rockets of the same mass. Regardless, they are very popular especially in semi-professional applications, such as in hobbyists' rockets or in engineering prototypes. The simplest hybrid rocket engine can be built almost completely from household materials using a stainless steel bottle for the combustion chamber, laughing gas as the oxidiser and a fatty salami sausage as the solid fuel!

A rocket vehicle is not limited to a single type of engine technology. In fact, multiple different types are often combined, as in the example of the Space Shuttle (see Fig. 2.16): The three liquid fuelled Space Shuttle Main Engines (SSME) at the back of the orbiter are actually producing the bulk of the thrust lifting the shuttle into space, even though their exhaust stream (a small blue flame of the oxygen-hydrogen reaction) looks far less impressive than that of the solid rocket boosters on the sides of the large main tank. While the solid booster's fuel is contained in the combustion chamber stretching throughout their white cylindrical shape, the engines on the orbiter are supplied with oxygen and hydrogen pumped from the orange external tank. Both boosters and the tank are jettisoned once in space.

Anecdote

One critical element of a liquid fuelled rocket is the injector, the part mixing oxidiser and fuel and pressing them into the combustion chamber (Fig. 2.13). It typically looks somewhat similar to a shower head from below, with alternating nozzles for the two fluid fuel components. Nowadays, computer simulations are used to design the injector to create an optimum blend of both components, ensuring smooth and steady combustion, but during the Apollo programme in the 1960s, this was not yet possible.

For the F1 engine of the Saturn V rocket, the injector was designed experimentally. Different prototypes were constructed, test run on an engine test stand and, due to the enormous power of this engine, they typically exploded immediately due to insufficient fuel mixing!

Finally, an engineer took a drill and bored a large number of holes into a metal plate by hand, which was then used for a further test. It worked flawlessly! From that moment on, all further F1 engines were built with an identical injector, where even the drilling mistakes were carefully duplicated.

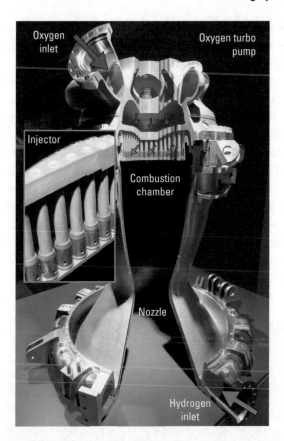

Fig. 2.13 Cutaway view of the Vulcain engine powering the first stage of the European Ariane 4 rocket. The oxygen turbo pump is directly mounted on top of the combustion chamber, while hydrogen first gets pumped around the nozzle and provides cooling before moving into the injector. There, the two fuel components mix before entering the combustion chamber (Photo from the special exhibition "Departure Into Space" at the Heinz Nixdorf MuseumsForum in Paderborn, Germany)

All rocket engines can experience combustion oscillations, when sound waves bounce back and forth inside the combustion chamber but they are usually easy to avoid in liquid and hybrid rockets since their chambers are relatively small. A more significant problem in these rockets is caused by *pogo oscillations* that have their origin in the fuel lines between the fuel tanks and the chamber. When the engine burns even a tiny bit inhomogeneously, it produces minor variations of thrust, making the fuel flow into the engine to vary ever so slightly. The variation of fuel flow can itself provoke the thrust to oscillate, inducing a feedback loop. If the engines are not designed to prevent this feedback, mechanical components around the engine can start to resonate with these oscillations, posing a danger of tearing the entire structure apart. Nowadays, the vibration frequencies are analysed and if necessary suppressed, to prevent pogo oscillations from causing any harm.

As a result of the combustion reaction, hot exhaust gases are created. Since higher temperature leads to better engine efficiency, the propellant mixture is chosen to burn as hot as possible, while making sure that the combustion chamber and nozzle walls do not melt or get ablated. Adding cooling mechanisms allows the combustion temperature, and thus efficiency, to be increased (unless the cooling mechanism adds so much extra weight that it eliminates the net performance gain). In liquid fuelled engines employing cryogenic (i.e. supercooled) propellants, the fuel itself can be pumped around the engine nozzle before entering the combustion chamber. This does not just cool the nozzle, but increases fuel pressure due to thermal expansion at the same time! Figure 2.14 shows an F1 engine of the Saturn V Rocket with the kerosene cooling pipes of the nozzle bell clearly visible.

Fig. 2.14 The F1 rocket engine powering the first stage of the Saturn V rocket is the most powerful liquid fuelled rocket engine ever created. The rightmost vertical pipes are fuel lines for liquid oxygen and kerosene that feed propellant into the turbo pumps. These, in turn, are driven by the precombustion stage (large cylinder at the top centre). At the bottom right, a white actuator arm is visible, which can pivot the entire engine to steer the rocket during ascent. The rings around the nozzle bell transport kerosene around it for cooling (Image: B. Ganse)

In solid fuel rocket engines and less complex liquid fuelled rockets, ablative cooling of the nozzle is a common approach. The nozzles are coated in materials deliberately chosen to cook away and take excess heat with them, similarly to the heat shields of atmospheric reentry vehicles (compare Sect. 2.2).

Independently of the fuel type used, after chemically reacting inside the combustion chamber, the resulting hot gases travel out through the nozzle at the stern of the engine. The higher their outflow speed, the more efficiently the rocket engine pushes the spacecraft forward. The shape of the nozzle, in which the combustion chamber first narrows down and subsequently rapidly expands into a bell shape, is called a Laval Nozzle (see Fig. 2.15). As gasses expand through the nozzle, they further push the rocket forward and gain velocity.

The ideal form of the bell-shaped outer part of the nozzle depends on the surrounding air pressure. For a high ambient pressure, the bell needs to be somewhat slimmer, whereas in a vacuum it has to be wider. When using an engine nozzle in an environment it has not been designed for, the gases flowing out will have too high or too low pressure, thus leading to overexpanded or

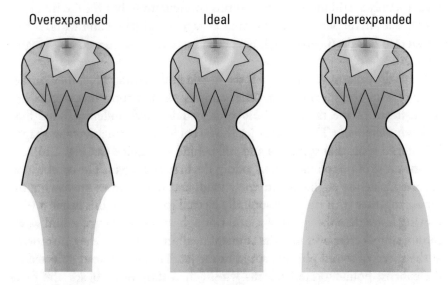

| Overexpanded | Ideal | Underexpanded |

Fig. 2.15 The shape of a rocket's exhaust stream changes depending on the surrounding pressure. The flow is called overexpanded if the nozzle form is too wide, and the gases will be pressed inwards by the surrounding air. In the underexpanded case, the gases still expand more after leaving the nozzle. In both cases, some of the thrust is wasted on sideways forces compared to the ideal operation mode, in which the exhaust stream continues straight out of the nozzle

underexpanded flow. Since the rocket's exhaust no longer streams straight out in those situations, but the gases' movement is somewhat sideways, the efficiency of the engine decreases. This is an additional reason for building multi-stage rockets. The simple rule is: The rocket engines of the first stage are optimised to operate in high ambient pressure, the engines of the further stages are set up for vacuum conditions.

The Soyuz rocket, even though technically a multi-stage rocket, is a notable exception to this rule. It carries a very long design legacy, dating all the way back to the Soviet Union's first intercontinental ballistic missiles in the 1950s. Even today, the engine configuration is still based on the old idea that starting a kerosene-oxygen engine in mid-air would be too unreliable. The Soyuz staging is hence built such that a central rocket core is surrounded by four boosters, such that the central core has a vacuum-optimised nozzle shape, and the boosters have nozzles designed for the dense atmosphere (see Fig. 3.1). All engines are ignited simultaneously before lift-off, and the centre engines are running at low throttle until the boosters are jettisoned. Even though better and more efficient staging methods have been developed long ago, the Soyuz design remains in use due to its unmatched reliability. The updated Soyuz-2 rocket family is still mostly based on this configuration, but for the first time contains a variant with a more modern 2-stage construction and without side boosters, called the Soyuz-2.1v. It has been in service for uncrewed cargo flights since 2013.

Once the exhaust gases have left the nozzle in the atmosphere, they further interact with the surrounding air. A curious phenomenon that occurs just downstream of the nozzle in both overexpanded and underexpanded flow is *Mach Diamonds*: The exhaust jet shows equidistant, roughly cone shaped brightened spots that seem to stand still behind the engine (see Fig. 2.16).

The cause of this phenomenon, oddly, lies in sound waves. The combustion process inside the engine is extremely loud, and sound waves are leaving the nozzle together with the supersonically ejected gases. Since this exhaust jet is moving faster than the speed of sound through its surroundings, sound waves cannot simply cross over into the ambient medium—after all, they are moving at the speed of sound plus the speed of the jet. They are instead reflected on the exhaust boundary and refocus within the centre of it. Inside the focus point, the combined energy of the sound waves heats the gases up to the point that they again start glowing hot. The same process repeats multiple times until the ejected gas has either slowed down to subsonic speeds or has become so turbulent that no common focus point forms anymore. Like sound, this phenomenon vanishes outside the atmosphere.

Fig. 2.16 Mach diamonds are visible in the exhaust plume of the Space Shuttle Main Engines (SSME) at the back of the orbiter. In these spots, shock waves focus within the engine exhaust. Since the engines are in use throughout the whole extent of the atmosphere, they are slightly overexpanded for low altitudes (Image: NASA)

Ignition

When a colossal rocket with impressive engines is standing on a launch pad, and a dangerously explosive fuel combination is supposed to be mixed and to react in a controlled combustion, one question arises: How is this system reliably ignited, so that all engines develop thrust at the same time, and no unburned fuel collects in or under the rocket? There are multiple solutions to this, differing in complexity.

Russian (and Soviet) space engineers are known to develop practical and sometimes ridiculous-seeming low-tech solutions to their problems. In the case of rocket ignition, they decided to go for a literal matchstick approach: Before launch, a simple wooden stick is inserted into the combustion chamber from below through the nozzle. A small pyrotechnical charge with an electric igniter is mounted at the top of the stick. All matchsticks are simultaneously ignited with an electrical pulse when the engines start. To verify successful ignition of the charges, a thin brass wire next to the charge has to melt away, and only when all wires have lost electrical contact, the fuel valves open and the rocket launches.

As simple and reliable this method may be, it can only be employed for a rocket ignition on the launch pad. Starting further rocket stages in mid-air requires different techniques. The situation is easiest when using hypergolic propellants, which ignite immediately on contact with each other, without the need for a spark or a pilot flame. Most of the available hypergolic fuel mixtures (and their combustion products) are unfortunately very toxic, typically involving hydrazine or one of its derivatives. While some of the Chinese Long March and the Russian Proton rockets still employ hydrazine-based fuel mixtures, they have fallen out of use for launchers in the Western countries completely due to health concerns. Once outside the atmosphere, they remain the fuel mixtures of choice for manoeuvring thrusters.

Even when the primary propellants are not hypergolic fluids, prefilling the engine tubing with such starter fluids is the preferred way to initiate the combustion. A less toxic mixture used both by the Saturn and SpaceX rockets combines 10% of triethylaluminium with 90% of triethylborane, or *TEA+TEB* for short. This starter fluid mixture burns with a bright green flame and causes the green flash seen in various rocket engines at ignition, for example, in the Falcon 9's Merlin engines.

Roll Control

Rockets launched from the ground are usually cylindrically shaped (or built as clusters of cylinders) for aerodynamic reasons. The engines are mounted at the bottom end and push the whole construction upwards, which results in a delicate unstable balance. So what prevents the rocket from toppling over and impacting the ground right after launch? In small hobbyist rockets and fireworks, aerodynamic stabilisation using small fins at the base, or even just an elongated wooden stick at the base of the rocket, are sufficient to counteract sideways motion. An active control system is required for any bigger design and whenever precise control of the craft's course is aimed for. It continuously measures the flight trajectory and counteracts any deviation from the desired flight plan. In the low and dense strata of the atmosphere, it can do so by actuating wings or control surfaces (which is advantageous, as it does not consume any propellant). These controls quickly lose their utility as the rocket zooms higher towards space. Actuating the rocket engines and thus steering by changing the exhaust direction is nowadays the most common method of rocket ascent control, and it works both inside the atmosphere and in a vacuum. It does, however, make the overall design of the propulsion system more complicated, as all tubes and conduits have to be built to be flexible, while still sustaining the same high pressure levels. If only one engine is available, roll control of the rocket can become a problem, since a single nozzle on the centre axis cannot really affect the roll direction just by tilting. Early versions of the Atlas rocket (Atlas 1–3) were therefore equipped with two additional small steerable rocket engines on the side of the main tank to enable roll control, working independently of the main engines at the bottom. Boeing's Delta IV and SpaceX's Falcon 1 rockets direct the exhaust of their turbo pumps for the same effect.

Soyuz rockets (and the R7 missile that it is based on) did not have any roll control for their initial design iterations. Apparently they were considered sufficiently roll-stable based on their shape alone, so no additional mechanism was considered necessary. Yet, also for a Soyuz craft, the flight direction needs to be variably selectable, so the entire launch pad construction was built to rotate into the desired direction. It was only with the introduction of the Soyuz-2.1a (in the year 2004), which can be launched from the European launch site in French Guiana, when roll control was added.

Small Engines for Space Usage

Fundamentally, rocket engines used in space work the same way as those used to launch from the Earth's surface. To start with, there is no need for them to be monstrously big anymore. Once the gravitational pull of Earth is overcome, there is no hurry to push with as much thrust as possible. Engines therefore tend to be small, and are rather optimised for fuel efficiency than maximum power. Furthermore, due to the fact that they are made for zero gravity, some technical details have their own peculiarities.

Control thrusters (Fig. 2.17; see more details of their function and handling in Sect. 3.2) are in the vast majority of cases built as liquid fuelled rockets. When trying to ignite them in zero gravity, their propellants cannot be assumed to collect "at the bottom" of a fuel tank—after all, there really is no "up" or "down" without gravity. Typically, fluids collect into a spherical blob to the centre of the tank and cannot directly be pumped out into a combustion chamber! So how do the propellants get from the tanks into the engine?

One option is to have a rubber membrane inside the (typically spherical) fuel tank, which keeps the fuel from sloshing around too much. As the tank empties, a pressurising gas (helium) gets injected through a valve on the opposite side of the membrane, and fuel gets squeezed out. The disadvantage of this method is its weight, as both the membrane and the helium need to be carried along, reducing the available payload mass. Another popular method of solving this problem is astoundingly similar to electronic cigarettes. Structures such as wicks (made out of glass fibre or a porous metal tissue) or sponges such as perforated metal sheets are submerged in the fluid and maintain contact with it due to surface tension, even as the propellant floats around the tank.

Fig. 2.17 Control thrusters, Vernier rockets and heat shield tiles at the stern of the Space Shuttle Atlantis (Image: B. Ganse)

Thanks to capillary forces, this *Propellant Acquisition Device* remains wet and draws the fluid towards the combustion chamber.

In the chamber, the fuel mixture and an ignition source need to be reliably brought into contact with each other, which causes yet an additional problem in zero gravity, since the movement of randomly sloshing fluid droplets in an injector is even harder to predict. The typical solution is again to use hypergolic propellants that spontaneously ignite without the need for an additional spark, but in contrast to the igniter fluids of big rocket engines, these now act as the primary fuels. For small thrusters, it can be sufficient to simply bring two fluid-carrying wicks into contact with each other to start the engine. Once they are separated, the combustion stops.

The exhaust of engines no longer looks like a normal flame once in vacuum and zero gravity, as it is missing the two important physical effects that would cause this shape: Convection is not occurring without gravity and air resistance has disappeared when the atmosphere was left behind. Instead, the slightly glowing gases simply dissipate rapidly outside of the engine nozzle.

In some cases, where hot exhaust gasses are undesirable, it is sufficient to have a compressed gas streaming out of the nozzle without any combustion. While this *cold gas thruster* design is not very efficient, it is uncomplicated (as it only involves opening a valve) and does not produce hot exhaust gases. Cold gas thrusters are therefore commonly found in manoeuvring packs of spacesuits.

Ion Propulsion and Plasma Jets

All engine types mentioned so far are based on chemical combustion—a chemical reaction produces energy, fuel material gets heated, expands and is ejected with high velocity. While this is a very successful method of ejecting mass with high velocity, it is far from the only one, and a number of additional methods have merit. In particular, alternatives include interesting engine concepts where at least part of the energy is supplied electrically, as solar energy is abundantly available in space. Therefore, the energy source for propulsion does not necessarily need to be added to the mass of the spacecraft.

The three most prevalent types of electrical rocket propulsion are (of which one is shown in Fig. 2.18):

- In an *arcjet* an electrical discharge takes the place of the flame. A gaseous propellant (such as nitrogen, hydrogen or ammonia) gets heated and expanded by the arc similarly as it would in combustion, but without

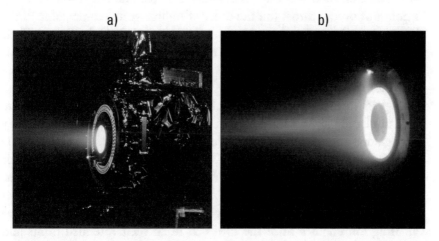

Fig. 2.18 Ion thrusters of two different designs. (a) NASA's NSTAR ion engine used in the Deep Space 1 probe. (b) A Hall effect thruster being test run inside a vacuum chamber. Both pictures show the round opening of the ion exhaust and the smaller tubular electron gun, which keeps the overall exhaust electrically neutral so that the spacecraft does not accumulate an electric charge (Images: NASA & Dstaack)

participating in a chemical reaction. The resulting ejection speed and efficiency are not tremendously better than in a chemical rocket, but an arcjet can avoid explosive or toxic propellants. Furthermore, an arcjet can be designed to operate inside an atmosphere, using ambient air as the reaction gas. All it would need is a sufficiently strong and lightweight source of electricity, such as a nuclear reactor. When powered by batteries or solar panels, however, they are limited to tiny amounts of force. Yet, as arcjets are technologically uncomplicated and can be built in miniature sizes, they are quite popular for propulsion in small satellites.

- *Ion engines* accelerate electrically charged particles in an electric field. All previous methods have attempted to produce fast atoms through thermal reactions, in which the atoms gain randomly directed kinetic energy and bounce into each other. When an electric field is instead used to give the atoms a unidirectional boost, a lot of inefficiencies can be avoided. The resulting outflow velocities can be close to the speed of light, thus offering phenomenal efficiency! The best choices of propellant are heavy but easily ionisable atoms e.g. mercury or xenon.

  Alas, there are two limiting factors that severely restrict the amount of thrust that can be achieved with ion engines: The rate at which ions can be produced and accelerated is limited by the available electrical power, while the density of the exhaust stream needs to be very low, so that atoms do not collide on their way out. As a result, these engines only produce a tiny amount of overall thrust (on the order of millinewtons, comparable to the force of an ant pushing the spacecraft), but are superbly fuel-efficient. Their primary utilisation is in manoeuvring of interplanetary space probes, for which it is sufficient to get only a small but continuous amount of force over multiple months.

- *Hall effect* thrusters are a subclass of ion engines, where the electrons are held in place by cylindrical magnetic fields while ions are heated until they have gained enough energy to escape. While more particles tend to hit the walls and electrodes of the engine, leading to increased abrasion of their materials, they are easier to build and create higher thrust than ion engines.

These engine designs are not just a vision of the future—spy satellites have been using ion engines since the 1960s to reach their intended orbits and observation targets. In civil use, the Russian *Meteor* weather satellites were the first to use ion thrusters for station-keeping. As of 2020, Hall effect thrusters are installed on the majority of new satellites. Among them are commercial geostationary satellites used for TV broadcasting that keep a constant position in the sky (Sect. 3.3 explains how they achieve this).

Nuclear Propulsion

While the established way to use solar panels or chemically stored energy as the power source for rocket propulsion works well, there is a form of energy production with massively higher yield per unit mass: nuclear energy. A nuclear reactor could be used to power the aforementioned electrical propulsion methods!

But why laboriously convert the energy release of nuclear reactions into electricity first, if the fast particles produced in these reactions could directly be used as a rocket exhaust? As it can be easily imagined, this train of thought was first picked up in the line of nuclear weapons research. Stanislav Ulam, known for the invention of the fusion bomb, proposed the *Orion drive* while working at the Los Alamos National Laboratory in 1946. He suggested building a spaceship carrying a round copper plate of multiple metres thickness at its aft side. Nuclear warheads wrapped in tungsten would be ejected from a central opening and detonated just a few metres behind the craft! The copper plate would act as a combined radiation and heat shield while also acting as the reaction mass, against which the ejected tungsten cover would push. Thanks to the absence of air in space, no high-density shock waves would form around the explosion, meaning that mechanical stress to the spaceship could actually be handled by shock absorbers between the plate and the ship.

The Orion drive is the only technically feasible spacecraft concept so far, which would in principle be able to fly an entire colony to Mars in one go. Calculations suggested that 800 tonnes of payload mass could be flown to Mars within 1 month ... , however, this would require that a nuclear bomb with 0.15 kilotons of explosive power were detonated behind the spacecraft every second! Already launching from the ground would have raised the entire Earth's background radiation level by about 5%. For obvious reasons, this concept has never been considered for implementation and most likely never will.

The *nuclear salt water rocket* presents a significantly less extreme approach to nuclear propulsion. Its fundamental design resembles that of a traditional liquid fuelled rocket engine, except that the propellant is formed of uranium, thorium or plutonium salts dissolved in water. Their concentration is chosen just so that no chain reaction starts while stored in the tanks out of neutron capturing materials like boron carbide. The reactant would be compressed and exceed the critical mass in the "nuclear combustion chamber" and start a nuclear fission reaction, releasing large amounts of energy. The evaporated water-fuel mixture leaves the engine as superheated steam pushing the rocket.

While this engine design is a lot less explosive and frightening than the Orion drive, it still requires weapons-grade nuclear fuels as the propellant. A launch accident of a rocket like this would be an enormous environmental catastrophe, so it will probably never be launched directly from Earth. Should the possibility arise to mine fissile materials straight from asteroids in the future, this rocket concept would be quite feasible for transportation from one distant celestial body to another. The radioactive products quickly dissipate into space, and any structure made for humans carries suitable radiation shielding already.

A more realistic nuclear propulsion design is the so-called NERVA engine (*Nuclear Engine for Rocket Vehicle Application*). In this engine, the combustion chamber of the rocket is replaced by a small nuclear reactor, into which propellant (for example, liquid hydrogen) gets pumped from the fuel tanks. When running at full power, this would heat hydrogen to much higher temperatures than a combustion process can, leading to increased expansion and higher nozzle exhaust velocities. In the 1970s, such an engine was actually fully developed and tested, with the intention to use it as the follow-up to the Saturn V rocket (called the Saturn C-5N). At the end of the Apollo programme, the development was discontinued for cost reasons.

Apart from nuclear fission, in which heavy atomic nuclei split, nuclear fusion, in which light nuclei combine to form heavier ones, (like inside of the Sun) is another potential method for propulsion. While the fusion products, helium or lithium, are still very light atoms and thus less efficient in transferring momentum, the power density of fusion is fantastic. To start fusion in the first place, the fuel needs to be heated to many million degrees though. Fusion energy for spacecraft propulsion remains an utopian dream, as maintaining fusion reactions with positive energy output has not yet even been achieved in laboratories on Earth.

Solar Sails, Elevators, Warp Drives and the Future

All propulsion methods discussed so far were based on the assumption that space is completely empty and provides no medium to react against, so that all craft dragged their reaction mass along in the form of fuel. But this assumption is not fully correct! Within the solar system the Sun is providing a steady stream of electromagnetic radiation (light), in addition to charged particles, called the solar wind. Both of these travel away from the Sun into deep space and carry an ever so slight momentum flux with them. If a structure were able to deflect this momentum flux over a sufficiently large area, it could be used to push a spacecraft, just like the wind pushes a sailing ship!

Photonic solar sails were already suggested in the 1970s. The idea was to unroll a sheet of vanishingly thin metal or mylar foil in space to use it as a mirror for sunlight. To produce any considerable amount of thrust, solar sails have to be many square kilometres in size and constructed as light as possible (Fig. 2.19a). Unrolling the sail has turned out to be logistically challenging, as thin metal foils in vacuum tend to stick together, because no oxide layers form on their surface. It is also complicated to keep the tenuous sail materials intact when performing manoeuvres. After many unsuccessful attempts to demonstrate this technology, both the Japanese IKAROS satellite and NASA's NanoSail-D successfully deployed small solar sails in 2010. While these experimental missions have shown the generation of tiny amounts of thrust, actual use of solar sail propulsion is still outstanding.

A different spin on solar sailing is given by the *electric* solar wind sail (Janhunen 2004). In this sail, the reflected particles are not the photons of sunlight, but the positively charged protons of the solar wind. To do so, the spacecraft itself needs to be positively charged. Charges of same sign electrostatically repel each other, so protons are pushed away once they get close enough to the surface of the craft. Compared to the photonic sail, this comes with the advantage that the sail is not required to be a continuous structure at all. A rotating spacecraft can fling out a set of wires and keep them tensioned with the help of the centrifugal force. Since the electrostatic repulsion force acts also between the wires, it gives an effective sail area shaped like a round disc (see Fig. 2.19b). As an additional bonus, these sails even allow

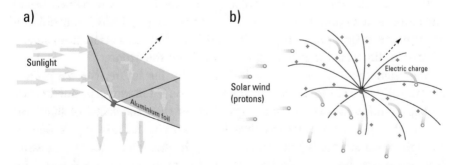

Fig. 2.19 Photonic and electric solar sail. **(a)** The photonic sail reflects sunlight off a large mirror area, thus gaining a (tiny) amount of net momentum. **(b)** The electric sail on the other hand creates a positive electric charge on the sail wires and spacecraft, in order to reflect protons of the solar wind

a limited degree of steering control, by varying the charge on the wires as they spin around.

In the end, both space sail technologies have one thing in common: They only yield ludicrously low amount of thrust, making them entirely unusable for human spaceflight. Nevertheless, they are a suitable choice for space probes on multi-year missions to other planets, or even for missions attempting to alter the flight path of asteroids that pose a future collision risk with Earth.

Completely different but no less unconventional is the idea behind the *space elevator*. As the name implies, this is a device to significantly facilitate and economise ascent through the atmosphere by building a giant elevator all the way up from the ground to orbit. This requires a fixed structure (basically a long cable) to stretch up beyond geostationary orbit (see Sect. 3.3), where a counterweight revolves along with the Earth in 24 h. The base of the elevator needs to be located on the equator to enable the top end to stay above the same spot. The centrifugal force of the counterweight keeps the cable tightened, so that a spacecraft can pull itself up using a simple electric motor. Even though this idea seems logical and practical, the actual construction brings some extraordinary difficulties. The material of the elevator wire needs to be both light and stable enough to hold its own weight over many thousand kilometres of length. No sufficiently strong material exists today that could be produced in these quantities (not even the strongest carbon nanotubes). And even if a suitable material were found: How would the construction take place? From the ground up, using a crane that has exactly the same problems? Or by reeling the elevator wire down from geostationary orbit? How would the ground station remain anchored? Further consideration of the problems of high wind strengths in the upper atmosphere and collision danger both with space trash and airplanes explains why this idea has remained utterly theoretical so far. Nonetheless, a space elevator is not physically impossible and might prove to be a promising method for launching spacecraft from smaller celestial bodies without an atmosphere, like from the Moon.

And then there is, as an idea for the really far future, the *warp drive*, scientifically referred to as the "Alcubierre metric". This concept purely exists on paper and would enable space travel faster than the speed of light! But how is that possible? Einstein's theory of relativity forbids any movement through space at speeds faster than light—the Alcubierre metric, however, does not move the spaceship itself at all. Rather, it describes a deformation of spacetime around the ship, where space contracts in front of it and expands behind it. The ship itself is sitting motionless in a spacetime bubble that does not contain any space curvature. Deforming space like this would require ridiculous amounts of energy far beyond the reach of mankind's capabilities. Furthermore, it demands some amount of negative energy density to cause spacetime to expand. No one has any idea how to create negative energy, so the warp drive will remain a set of equations on a piece of paper for the foreseeable future.

Attitude and Position Determination

Being able to effectively manoeuvre a spacecraft does not just require one of the previously discussed propulsion methods, but it is also important to have an idea about position and orientation (or attitude) of the craft. In the first spaceflight Yuri Gagarin still oriented his spacecraft by hand and eye and estimated the appropriate moment to fire the reentry rockets by looking for landmarks over South America. Nowadays a sophisticated combination of systems work together to provide guidance, navigation and control (GNC) capabilities.

The central part of this setup is the navigation computer, which combines sensor data to provide continuously updated information about position, velocity, direction and rotation of the vessel. This information in its entirety is called the *state vector* of the spacecraft.

While the computer is able to update the state vector by simulating the spacecraft's physics even without measurement inputs, it will increasingly drift away from the actual situation of the craft due to numerical inaccuracy and unforeseeable physical effects (like crew members moving around inside the vessel or micrometeoroids hitting the hull). If the state vector disagrees with sensor measurements, the difference has to be fed back into the simulation to keep accuracy sufficiently high. For automatic spacecraft docking procedures, for example, the position needs to be known within centimetre precision!

Several kinds of sensor technologies are available to gain information about spacecraft attitude and position:

A gyroscope is a spinning mass that is mounted to be freely rotatable in space and is used for determination of spacecraft attitude. If the spacecraft rotates in any direction, the central spinning mass tries to retain its spin direction. This causes a torque to act on the gimbal axes around it (Fig. 2.20a), moving them in relation to the new spacecraft direction. No external reference point is

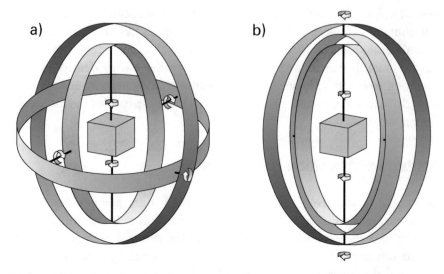

a)   b)

Fig. 2.20 Gyroscopes are used to determine orientation in space. (**a**) A spinning mass mounted in the centre rotates the rings surrounding it when the assembly is tilted. (**b**) In a gyroscope with only 3 axes, a situation can occur where multiple axes are pointing into the same direction, making the instrument inoperative (the so-called gimbal lock)

required for it to work. Over time, however, gyroscopes drift out of alignment due to rotational forces (the same effect that makes a spinning top wobble on a table). Occasional recalibration from other reference sources is therefore necessary.

Spinning masses not only act to determine the attitude, but can also be employed to actively control it, in *control moment gyroscopes* (CMGs). They provide a way to orient spacecraft without the use of thrusters, by instead exchanging momentum between a spinning mass and the spacecraft. Structurally, they are very similar to navigational gyroscopes, except that the mass spinning in them is much larger, and the motors turning the axes are stronger. Space telescopes and observational satellites especially rely on them for attitude control, to minimise dirt on their optics from thruster exhaust gases.

Excursion

The Apollo missions primarily relied on a gyroscope to measure spacecraft attitude for the guidance computer to make navigation decisions. In order to arbitrarily turn in space, it was fitted with a 3-axis mount. This kind of system has a problem, however: Any 3-axis mounted gyroscope system has a pair of orientations, in which two of the axes align with each other, causing the system to jam (a situation referred to as "gimbal lock", see Fig. 2.20b). The Apollo gyroscope was deliberately constructed in a way that the critical directions would align with the north-south axis of the Earth, as it was assumed that the spacecraft's nose would never have to point exactly north-southward during the flight to the Moon and back. Otherwise, the gyroscope would have locked, and the navigation computer would have become inoperative. A complete restart and recalibration of the attitude system would have been necessary.

During the Apollo 13 incident an oxygen tank exploded on board the spacecraft. It sent the vessel into an uncontrolled tumbling motion, in which the orientation came dangerously close to the gimbal lock direction and the computer triggered an alarm sound. Luckily, control was regained and the attitude was stabilised in time.

This problem is normally circumvented by adding a fourth axis to the gimbal platform which is able to turn the whole system away from "dangerous" orientations. Back then, when the Apollo spacecraft was developed, this was considered too heavy, too complicated and too expensive. The astronauts were well aware of this situation: Michael Collins asked for a fourth gyroscope axis for Christmas during the Apollo 11 mission.

Fixed stars in the sky are, as the name implies, fixed (at least on the timescales of a human lifetime). They are therefore perfectly suited to determine spacecraft direction. In early spaceflights up to the Apollo programme, spacecraft were equipped with navigational telescopes that were used to get the bearings of certain reference stars. Reading the bearing angles and feeding them into the navigation computer compensated for small inaccuracies that the gyroscopes would accumulate over time. Nowadays, the same process is done automatically using *star trackers*. These are nothing else but digital cameras, which compare the pictures they take with a database of bright stars, thus determining the attitude. Multiple cameras are usually mounted facing different directions, so that their functioning can also be ensured when one side of the craft is shadowed by Earth or bright sunlight is glaring into them. One important limitation of star trackers is that they require a sufficiently stable alignment to take their pictures without motion blur and will not work when the spacecraft is rotating or tumbling quickly!

Excursion

The Rosetta mission to the comet 67P/Churyumov–Gerasimenko relied on star trackers for attitude control. As the probe was approaching the comet, it suddenly started performing unpredictable manoeuvres by itself!

As it turned out, the comet was heating up while approaching the Sun, and small dust and ice particles released from the surface were drifting past the probe at slow speed, crossing the fields of view of the star trackers. Since these white dots looked like moving stars to the cameras, the navigation computer determined that it must be unexpectedly rotating and tried to compensate for it. For a while, Rosetta was therefore commanded to keep a bigger safety distance from the comet until a software update for the star trackers was developed to better differentiate stars from dust.

It is easy to determine at least one direction in space using the Sun as a reference light source. Tracking the Sun does not even require a particularly sophisticated camera. A handful of photodiodes scattered on the surface of a spacecraft is entirely sufficient for a rough determination of orientation. Likewise, horizon trackers can give some information about the angle between the spacecraft and the horizon of the body it is orbiting by looking at light scattered from the body.

The previously mentioned systems have been able to measure a spacecraft's attitude, but provide no information about its position. This is a notably more difficult process. The easiest partial position information when orbiting a planet is the altitude of the flight path. By measuring the apparent radius of the planet (observable as the angle from horizon to horizon), the height can easily be calculated from the planet's actual diameter with a bit of trigonometry. Radar altimeters give even more precise height information in the case of moons or planets without an atmosphere.

Identifying which point on the surface a spacecraft is flying over means finding longitude and latitude. From the vantage point of the spacecraft, this is more difficult than finding the altitude, especially when easily identifiable geologic features like mountain ranges or craters are regularly covered by clouds, like on Earth or Titan. It is much easier the other way around: Spacecraft in orbit can easily be detected and tracked from the ground, as anyone can confirm who has seen the International Space Station whizz overhead in the evening (the QR code in Fig. 2.21 links to a website predicting satellite overpasses that are visible with the unaided eye). No matter whether tracking occurs optically or via radar, the information can be uplinked to the spaceship through a ground station. This provides complete, three-dimensional position information when combined with a height measurement.

Fig. 2.21 The website Heavens Above (http://heavens-above.com) contains a huge catalogue of satellites and can calculate overflight times that are visible to the naked eye from any given location on Earth

Another present-day technique to determine spacecraft position is a satellite navigation system—the same kind that is used for navigation in phones on Earth. In principle, these navigation systems, such as GPS, Glonass or Galileo, are designed to be used on or near the ground, with velocities much smaller than the orbital speed of the satellites. A receiver compares the signal arrival times from different satellites and calculates the position relative to them. The same method can be applied when orbiting far above the navigation satellite orbits, even though it is mathematically more difficult to pinpoint an exact location. The amateur radio satellite Oscar 40 was the first to demonstrate a successful GPS fix from 58,700 km altitude in the year 2000. This has since become a standard method for position determination in low Earth orbit. Also the newest generation of Soyuz spacecraft, called Soyuz MS, can determine their orbital position using a combination of GPS and Glonass satellite signals.

2.4 Power Supply and Temperature Regulation

The first human spaceflights were completely relying on batteries for their electricity supply and, due to their short flight times, did not need to worry too much about temperature regulation. As soon as the spaceflight duration exceeds a couple of hours, power supply and thermal regulation become important concerns.

How to Produce Energy

On Earth, the typical process of energy generation involves spinning a generator wheel. This rotation can come from steam generators, water power or wind power. Rocket engines can, as a by-product of their thrust, also supply copious amounts of electrical energy, for example, by driving generators on the same axis as the turbo pumps (compare the beginning of Sect. 2.3). Once a stable orbit has been reached, and the engines are shut down, other energy sources are needed before the batteries run out.

Once out of the atmosphere with its clouds, dust particles and ozone layer, nothing blocks sunlight from reaching the spacecraft anymore. Sunlight in space contains a larger fraction of shorter wavelengths (UV light), which get filtered out in Earth's atmosphere. Since shorter wavelengths carry more energy per photon, the efficiency of solar panels in space is even higher than on the ground. No wonder that they have been the energy source of choice throughout most of the Space Age. The American satellite Vanguard 1, launched in the year 1958, was the first to use solar cells for power supply. The International Space Station is equipped with vast solar arrays that deliver a nominal power of 120 kW while in sunlight. This corresponds to about one tenth of the power generated by a modern wind turbine or one ten-thousandth of a nuclear power station's output and is significantly higher than the power consumption of an average family home. Of course, the spacestation uses electricity for a lot of additional functions that a house on Earth does not need, such as air processing, attitude control and scientific experiments (Fig. 2.22).

Use of solar cells in space comes with two important phenomena that require attention in the design:

- Electric charges build up in certain spots because charged particles, in form of cosmic rays and solar wind, impinge on the surfaces of the spacecraft. In areas where the hull material is electrically conducting (for example, built from metal) these charges quickly neutralise and do not pose a problem. On the glass surfaces of solar panels, however, surface charges can accumulate hundreds of volts electrical potential. This can affect the electron motion inside the solar cells and decrease their efficiency or even damage them, if a sudden discharge occurs after reaching a breakthrough voltage. Unfortunately, this surface charge cannot easily be used for power generation, since the overall energy is flimsy—charges build up over a couple of days, and immediately collapse once an electrical circuit is formed. Whether or not spacecraft charging will be able to contribute to the power supply in the future is still being investigated.
- Cosmic radiation also leads to changes in the semiconductor materials of the solar cells, as atoms are "kicked" out of their crystal lattice sites. This results in a gradual loss of power output over time. The rule of thumb is that silicone-based solar cells lose about 30% of their peak power over 10

Fig. 2.22 Solar arrays of the International Space Station. These 35 m long arrays are made out of solar panels, which themselves consist of individual cells. The entire array assembly rotates, so that it directly faces the Sun for optimum efficiency (Image: NASA)

years in low Earth orbit. These changes are irreversible, so solar panels have to be replaced after a couple of decades, and the complete electrical system needs to be planned for a steady loss of input power.

Solar cells work wonderfully and reliably to supply electric power near Earth and missions to the inner planets. The situation looks different in the outer solar system, or beyond the orbit of Jupiter. Since solar irradiance decreases further away from the Sun, eventually the power output of solar panels gets so low that they are no longer useful in relation to their weight. The same is true for spaceships or stations spending extended periods of time in the shadow of a celestial body, e.g. stations on the surface of the Moon.

A further, relatively compact and time-tested method of electricity generation is the *fuel cell*. It is an electrochemical reaction chamber, in which hydrogen and oxygen combine at a thin membrane to form water. This is basically the same process as in the combustion of these two propellants inside a rocket engine, except that the same reaction equation takes place in a fuel cell without producing a flame. The membrane (e.g. a Nafion foil) allows protons to pass through but blocks electrons and causes them to "take a detour" through the electrical circuit, powering it. Altogether, this turns the chemical reaction energy into electrical current. The cell itself can be quite small (no larger than a mobile phone), and the energy density of the propellant lies an order of magnitude above lithium ion batteries! Fuel cells are a remarkably practical energy source for spacecraft that carry oxygen and hydrogen anyway as rocket propellants. They were extensively used during the Apollo missions, powering the entire command and lunar module.

Even the energy amount that oxygen and hydrogen can carry is not sufficient for long duration missions lasting many years. Therefore, the power supply of space probes on missions to the outer planets has to rely on nuclear energy sources for now.

Radioisotope batteries, also known as Radioisotope Thermal Generators (RTG), are the least complex form of nuclear energy. Inside them, a long-lived radioisotope, like plutonium 238, disintegrates and heats up due to the decay energy release. The temperature difference between the isotope and cold space (or the cooling system of the spacecraft) can be used by a thermoelement to create electric power. While the overall power output of such a battery only lies in the order of a few watts, plutonium 238's half-life of 87 years allows them to be operated for a century. Figure 2.23 shows one of the plutonium pellets that powered the Cassini space probe as it explored Saturn and its moons. Generators of similar design have been installed in the Voyager probes, the Galileo spacecraft and the Mars rover Curiosity.

Fig. 2.23 A pellet of plutonium 238 dioxide, glowing red due to its own decay heat. This pellet was used for the radioisotope batteries of the Cassini space probe. The temperature difference between the plutonium and the cooling system of the spacecraft can be used to generate electricity (Image: NASA)

Another option for long-term energy supply with much higher power output is a nuclear reactor, where the energy production rate can be actively controlled and regulated. In uncrewed spacecraft, these can be run with minimal shielding. Only electronic components need to be protected against radiation, with the same techniques that double as shelter from cosmic rays (see Sect. 5.7). Space-borne nuclear reactors are thus notably lighter than the ones used on Earth. However, using nuclear reactors on spacecraft is immensely unpopular because of the negative image of nuclear power and the threat of contamination in the case of a launch failure. Only a single American mission, called SNAP-10A, tested this technology, whereas the Soviet Union equipped a total of 33 satellites with a multitude of nuclear reactor models. Most of these satellites had secret missions and were most likely used as spy satellites, also evidenced by the fact that they were painted completely black.

Hamsters running in spinning wheels are unfortunately not suitable for energy generation in zero gravity environments, as the hamsters would not be able to hold on to the wheel and would drift away. If the wheel itself were spun to provide artificial gravity for the hamster, more energy would be consumed to keep it spinning than the hamster would produce. Hence, it would no longer work as an energy source.[2]

Temperature Control and Radiators

As already mentioned in Sect. 2.2, there is a significant temperature difference between the sun-facing side of the spacecraft and the one in the shadow. The vacuum of space in addition acts like a thermos bottle, preventing any thermal conductance so that any heat produced inside the craft will stay in it. So how can temperature be managed in space, if no surrounding medium is available to shed heat into?

Since the vacuum of space has zero pressure, any drop of liquid has higher internal pressure and vaporises. One straightforward option is to bring the cooling medium along: A liquid, typically water or ammonia, is heated in a heat exchanger and exposed to the vacuum. As it evaporates off into space, it takes heat with it, thus cooling the vessel. The Space Shuttles employed two different "Flash Evaporator" systems: a water-based system for vacuum use (with a high heat capacity, carrying away plenty of heat per kilogram) and an ammonia-based system for the atmosphere below an altitude of 32 km (with a lower boiling point, so that it still cools where water does no longer evaporate

---

[2]The authors are welcoming experimental verification of this claim from space.

effectively). This naturally consumes the coolant fluid over time. Thus, their use is typically limited to the launch and landing phases, in which other cooling methods are impractical.

For more permanent space habitats, cooling systems need to work without consumable substances and rely entirely on infrared radiation cooling. Surfaces on the dark, cold, non-sun-facing side have a net heat loss as they are giving off more infrared light than they receive.

Figure 2.24 shows an example of cooling panels in the International Space Station. The station's cooling cycle is also working with ammonia because of its boiling point well below room temperature. As opposed to the evaporative system mentioned before, it circulates in a closed loop from which none gets evaporated into space. The ammonia runs through the cooling panels, allowing them to radiate part of the heat away. This system is only able to cool from "hot" to "warm", since radiation cooling becomes more efficient the higher the temperature difference. The water taps on the spacestation thus only offer the choice of hot or lukewarm water for everyday use. Cold water for scientific experiments is separately cooled in small quantities using a significantly less energy efficient refrigeration system.

Fig. 2.24 Cooling panels (white) on the International Space Station. The ammonia coolant cycle of the station flows through these panels which radiate heat out into space as infrared radiation (Image: NASA)

2.5   Life Support Systems

Breathing Air

Human beings require oxygen to live. On the Earth, air contains about 21% oxygen, the other constituents being 78% nitrogen and 1% other gases (among them argon and neon, without relevance to the human body). Strong deviations of the oxygen content in either direction are bad news: An oxygen level below 15% leads to undersupply (hypoxia), while considerably higher concentrations than normal cause toxicity and increased cellular ageing (see Chaps. 5.5 and 6.4 for more details about humans' breathing requirements).

Beyond its biological effects, cabin air composition influences further properties and functions of the spacecraft. The oxygen content is directly related to the fire and explosion risk on board while air moisture affects electrical circuitry and corrosion.

Maintaining a breathable air environment includes two basic processes. The depleted oxygen content needs to be replenished in the cabin air and the breathing product carbon dioxide, CO_2, has to be removed from the air mixture, all while keeping the cabin pressure at a constant level. There are three main methods to supply oxygen inside spacecraft:

- The easiest way to get air is—like in scuba diving—from compressed air bottles. Air is directly delivered to the spacefarers via a pressure regulator, and the exhalation is simply vented out to space through a valve. This is light and uncomplicated in its construction, but also needlessly inefficient: The exhalation of a human being still contains about 14% of useable oxygen that could in principle be recycled into the breathing system. This method was appropriate for the first short spaceflights, but current spacecraft use a closed cycle, in which air is reprocessed and reused.
- A second option for oxygen supply is an *oxygen candle*. These typically contain a mixture of iron powder and lithium chlorate or sodium chlorate, which is heated up to about 600 °C. The chemical reaction between these substances forms sodium chloride (regular salt) and iron oxide (also known as rust), with oxygen as a reaction by-product. Once activated, each unit gives off oxygen slowly, overall producing 600 l, which covers the average daily consumption of a male crew member. Since they are reasonably compact and technologically uncomplicated, they tend to be used as a fallback system when the primary life support systems are out of order. Once

their chemical reaction has run to completion, they are discarded, as they cannot be reverted to their original state.

- Oxygen generation is relatively easy, if electric power and water are available. Simply by inserting electrodes into water and running an electric current between them, the water molecules get split up into hydrogen and oxygen. The water used for this purpose does not even have to be particularly pure or valuable drinking water, but filtered or distilled waste water will do the job just as well. Both the Russian Elektron system and the Environmental Control and Life Support System (ECLSS) in the American part of the International Space Station use this method to supply oxygen to the cabin air.

The second important task in breathing air processing is to remove the carbon dioxide (CO_2) that humans exhale. The CO_2 content in Earth's atmosphere is nowadays a bit over 0.04% (400 parts per million). Already when the CO_2 level reaches 1%, tiredness and difficulty to concentrate occur. At 7–10% suffocation symptoms and unconsciousness set in. The device performing the task of carbon dioxide removal is called a CO_2 scrubber and can be built in multiple ways:

- When CO_2 gets in contact with water, it forms carbonic acid, like noticeably present in a carbonated beverage. Like any acid, it can be neutralised by an alkaline base such as potassium hydroxide (caustic potash) or lithium hydroxide. The combination of these two reactions effectively reduces the CO_2 content in the air. For this method to efficiently work in zero gravity, the contact area between cabin air and the alkaline water solution needs to be as large as possible. This can be achieved by using mesh-like surfaces on which this chemical interaction takes place. The filters themselves are single-use items that are saturated and have to be exchanged after a couple of hours to days. They are, however, very reliable and pack compactly. They are thus used for small spacecraft and most prominently in the life support packs of spacesuits.
- CO_2 molecules are readily adsorbed to certain surfaces, such as charcoal, metal oxides or zeolites (porous minerals). In this process, a single-molecule layer forms on the filter until it is completely "covered". This works particularly well with materials that are textured with a large amount of very small pores, as their effective surface area is greatly increased. Simply running the cabin air mixture over such a surface binds the CO_2 to it, while oxygen and nitrogen just stream past. After only a couple of minutes, however, the surface is saturated and will stop collecting further carbon

dioxide. Once this has happened, the filter is isolated from the cabin air by closing a valve and subsequently exposed to the vacuum of space. By heating it up (in the case of charcoal, to about 200 °C), the bound CO_2 molecules are released, and they vaporise away, thus allowing full reuse of the filter surface for the next such cycle.

Aboard the International Space Station, the Russian Vozdukh system, the European Advanced Closed Loop System (ACLS) and the American Carbon Dioxide Removal Assembly operate with this principle, by automatically cycling multiple filter containers between CO_2 adsorption and vaporisation.

- The *Sabatier* reaction is a particularly useful piece of chemistry. In this reaction, carbon dioxide (CO_2) and hydrogen (H_2) form methane (CH_4) and water (H_2O). The two input chemicals need to be mixed, heated to temperatures of 300–400 °C and streamed over a nickel or ruthenium catalyst. Both of the resulting products are useful in spaceflight, as the water can directly be fed into the drinking water system, and methane is suitable as a rocket propellant. An experimental air processing system with this mechanism has been working on the International Space Station since 2010 and a second one was installed in 2018 with the European ACLS. The same reaction could be used to refine the atmosphere of Mars, with its high CO_2 content, into rocket fuel for flights returning to Earth.

- Just like on Earth, plants and other organisms can remove CO_2 from the air through photosynthesis. In their green parts, biochemical reactions convert CO_2 to sugar, feeding the plant. As an added benefit, oxygen emanates into the environment. Photosynthesis requires a suitable plant growth environment, containing plenty of water and the correct wavelengths of light, which takes up a considerable amount of space.

  The first life support system using biomass is being tested on the International Space Station. The Photobioreactor experiment houses a small amount of *Chlorella vulgaris* algae with some added nutrients in a water tube, through which CO_2 enriched air is run from the CO_2 exhaust port of the ACLS. The algae produce oxygen through photosynthesis, binding carbon into their own biomass and it can become a food source for the crew.

  In future space colonies and settlements on other planets, plant-based systems are expected to be of central importance (compare Sect. 6.5).

  A further option to bind CO_2 is offered by chemosynthesis, which is used by some microscopic organisms to fuel their metabolism. Some species can remove CO_2 from the environment without needing light as an energy source. Instead, energy is supplied chemically through a redox reaction, whose substances have to be brought along. Also, these organisms usually thrive in anaerobic and dark environments quite uncomfortable to human life, such as hydrothermal vents on the ocean floor, where they do not have to contend against photosynthetic competitors. At the moment, chemosynthesis is not of keen interest for spaceflight but it might find use in situations where sunlight is scarce but a suitable chemical energy source is readily available, such as the night sides of celestial bodies.

Excursion

On the way to the Moon, an oxygen tank exploded on board the Apollo 13 spacecraft due to a faulty heating coil (it was supposed to be isolated, but liquid oxygen leaked into it). The explosion damaged or destroyed many of the systems of the command module, including the fuel cell power supply. In order to survive the trip back to Earth, the three-person crew had to depend on the systems of the lunar lander module, which unfortunately were only designed to support two people for 2 days. Its CO_2 absorber canisters (Fig. 2.25a) were fully saturated after a day, and the carbon dioxide content of the cabin air rose to dangerous levels. The canisters had to be replaced by spares from the command module—except that module had been built by an entirely different company, and the filters had an incompatible shape (Fig. 2.25b).

Finally, an adapter contraption (Fig. 2.25c) was improvised using a multitude of items that could be found on board, among them piping and air pumps from a spacesuit, plastic bags, socks and a large amount of duct tape. It worked, and the astronauts returned home safely.

a)   b)   c)

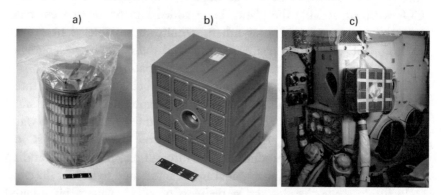

Fig. 2.25 Canisters containing lithium hydroxide are used to absorb carbon dioxide from the air. These models were used during the Apollo programme. **(a)** Canister for the lunar lander module. **(b)** Canister for the command module. **(c)** During the Apollo 13 mission, an improvised adapter had to be built to fit command module canisters into the lunar lander module (Image: NASA)

Apart from supply and removal of breathing gases, another important function of the life support system is the filtering of smells from the cabin air. In older spacestations, up to the Mir (1986–2001), smell was not considered an important quality, which led to the development of a pungent odour inside the station over its 15-year lifetime. During its last operational years, bacterial and fungal infestations in the station's air treatment system made the smell increasingly unbearable. Cosmonauts reported extended episodes of nausea on arrival to the station or when returning from spacewalks, where they had been breathing clean air for a short period of time. Thankfully, lessons were learned from Mir's smell developments and have been taken into account on the ISS. Its life support systems are designed with antibacterial surface materials as well as replaceable air filtration assemblies, to keep unwanted odours away.

Nonetheless, the International Space Station is said to have a very characteristic, although not unpleasant, smell of its own. Whenever a new crew arrives to the station and opens the spacecraft's docking port, they are remarking on the "typical ISS smell", which is strange and metallic. It has been said that this odour cannot be sufficiently described in terms of smells known from Earth. It might be one of the most exclusive sensual impressions available to human beings.

Water

Water is a further essential resource for the subsistence of human life, both in liquid form and as moisture in air. To be suitable for human consumption, it needs to be free of harmful microbes and chemical contaminants. For short spaceflights, water can simply be brought along, packaged like food supplies. Longer missions and spacestations demand water recycling systems.

Wastewater inside a spacecraft can stem from several different sources. Some scientific experiments can have water returns, air moisture is condensed into liquid form and most importantly, the liquids obtained from the spacefarers' bathroom have an overall water content that matches the requirements to cater the crew.

Wastewater is processed in multiple stages to turn it back into a usable product. First, the water gets filtered to remove any solid or gaseous impurities. Next, it gets pumped into a reaction chamber where a catalyst surface eliminates any remaining organic compounds. The vast majority of remaining trace substances are removed by distillation, i.e. boiling and condensing the water in multiple stages. Since distillation requires gravity to work, it happens in a centrifuge. In Fig. 2.26, the centrifuge of the International Space Station

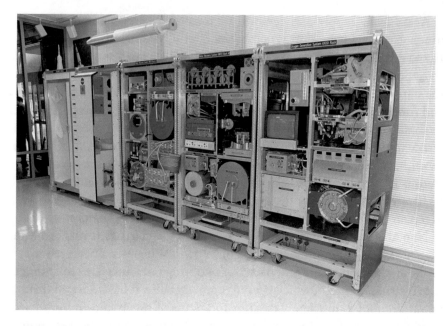

Fig. 2.26 The Environmental Control and Life Support System (ECLSS) of the International Space Station (shown here: a mock-up model on the ground), consisting of oxygen supply, CO_2 removal, air filtration and water processing systems. Each rack is 2 m high and 1 m wide (Image: NASA)

life support system is visible as the cylindrical object with many tubes attached, in the very bottom of the third rack from the right.

The output water is analysed by an automatic measurement system that decides whether or not it is suitable for consumption as drinking water. Otherwise it recirculates through the same process once again. Altogether, the water recovery system on the ISS is designed to have about 85% efficiency. Around one litre of water gets removed from the reuse cycle per day, hence water is delivered in bulk with transport spacecraft to refill the supplies.

The majority of the recycled water is not used for drinking, but fed into the oxygen supply system, where it gets split into hydrogen and oxygen (compare the beginning of Sect. 2.5). A common question asked from astronauts is, if the water recycling system on the International Space Station forces them to essentially drink their own urine, can thus be clearly answered: No, it forces them to breathe it!

The Bathroom

Water does not only serve as a critical substance for supporting human life, it also has important applications for overall wellbeing and hygiene, through bathing and washing. But how does bathing work in weightlessness? Using a bathtub filled with water is completely impossible, since water would have no reason to stay inside. It would float around the spacecraft's interior and even pose an electric shortcut risk or a suffocation hazard.

The same is true for a sink—like in the bathtub, the water would have no reason to stay put and would form a blob in mid-air. While water taps exist on the water processing system, their primary function is to fill small containers. To drink, cook and wash, water is usually first put into plastic bags fitted with a straw. Water is then squeezed out of them in small amounts, as required.

A shower, on the other hand, has been made to work without gravity! The American Skylab spacestation in the 1970s was the first to include it. There, water was sprayed out of a shower head on one end and collected by vacuum suction at the other end. While both the suction mechanism and the process of drying off inside the cabin were plagued with complications, a shower cabin based on a similar design was developed for the International Space Station (the leftmost unit in Fig. 2.26). Its mass, size and water consumption were in the end considered unreasonably large, so it was never actually delivered to the station.

Hence, in the absence of a shower, only wet rags are available as bathing equipment on the ISS. Since water is bound to them by surface tension and does not drip away without gravity, these rags can transport a surprising amount of water around without filling the inside of the station with droplets and bubbles.

The last essential bathroom item is the toilet. Up to and including the Apollo missions, there was only rudimentary toilet infrastructure available in American spacecraft: a pipe was used for urine collection (which was vented into space), while solid waste was collected with plastic bags that were taped to the buttocks. Astronauts disliked the facilities so much that they deliberately ate less during their missions, to reduce the need for toilet breaks. Russian technology was ahead of the American one here, as the Soyuz spacecraft contained a functional toilet assembly starting from 1967, which still involved manual handling of filter bags for separation.

Progress has been made and modern space toilets (Fig. 2.27) have become quite similar in usage to the ones on Earth. Anybody who has used an airplane toilet knows that they use only tiny amounts of water compared to those at

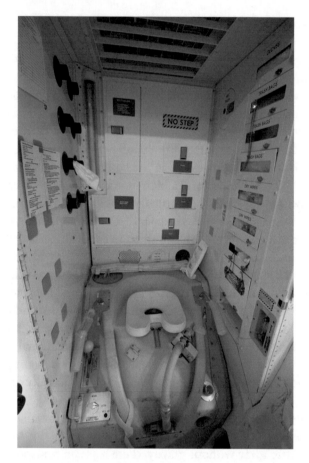

Fig. 2.27 A Space Shuttle toilet (in the Johnson Space Center). Orbital outhouses such as this usually work with vacuum suction (Image: B. Ganse)

home. Rather than flushing with water, vacuum suction empties the bowl into a septic tank. A similar process works in space, where a steady airflow replaces the pull of gravity. The septic tank occasionally gets replaced and handled like any other waste item. On the International Space Station, both toilets in the Zvezda and Tranquility modules currently have a Russian design (a new American toilet is awaiting delivery). The ISS tour video (see QR code in Fig. 4.3) by the American astronaut Sunita Williams explains the usage and functions of the toilet.

Redundant Systems

This chapter has already addressed redundancy and failure safety in most of the systems, structures and technologies described. Since technology is fundamental for survival in space, any critical component needs to be either repairable or swappable in flight if it malfunctions.

Systems inside the craft, such as life support and computers, can in principle be serviced by the crew (provided that spare parts are available), whereas rocket engines, airlocks, landing legs and the hull are hard to access and repair while in space. During certain stages of a spaceflight, such as launch or reentry, repairs are even more difficult due to time constraints and environmental conditions. Hence, all essential functions need to be failsafe and the easiest way to achieve this is to have multiple redundant systems fulfilling the same function.

The rocket engines used at launch are a good example of redundancy. All spacecraft are designed so that one or even multiple engines can fail without endangering flight orientation or causing an immediate crash—and yet they carry launch escape systems. Control thrusters are arranged in groups, so that control is retained even in the case of one thruster failing. Life support systems typically have a backup available that fulfils their function for at least a limited amount of time.

Whenever possible, it is recommended to build redundant systems with different operating principles, or to have them built by separate companies. If a systematic manufacturing defect were to appear in both the primary and the backup system at the same time, all the redundancy would go to waste! However, one is well advised to retain compatibility of parts, as the story of the Apollo 13 CO_2 scrubbers illustrates.

During the Apollo era in general, some design decisions were deliberately made against redundancy to conserve weight. In the Apollo lunar lander module, for example, neither the life support system nor the rocket engine of the ascent stage had any built-in redundancy. A broken-off switch in the instrument panel of Apollo 11 almost caused the astronauts to be stranded on the surface of the Moon, had they not fixed it using a pen. Overall, the Apollo programme was run on a razor's edge in terms of component failure, and could have easily turned into a big catastrophe, instead of the "giant leap" that it is celebrated for. Future missions to distant celestial bodies will hopefully be built to higher reliability and redundancy standards.

2.6   Communication, Computers and Data Management

While spaceflight is spectacular and enjoyable, it can get very lonely in the vastness of space. To keep a connection with the world outside the spaceship, radio communication and a system for handling and transferring data are needed.

Radio Connections

Any spaceflight requires some ground infrastructure and radio transmission planning before launch, as normal mobile phones (up to the 4G standard) will not work in space.[3]

Since the distances in space can get pretty big, good and highly directional antennas are recommended on both ends of the radio link. It would be ideal to have an antenna dish which focuses the majority of transmitted power towards the target, because otherwise a large part of the signal would dissipate away into empty space. But be aware: It is a good idea to have a second, less directional antenna on board, so a connection can be maintained even when not pointing straight earthwards, for example, when the spacecraft is tumbling or turned into a different orientation for whatever reason.

Compared to radio connections between two stations on Earth, ground-to-space radio links do not have a lot of things obstructing their path. The atmosphere stops having an effect on radio propagation above the height of the ionosphere (60–220 km altitude) and any other barriers, such as mountains, trees or buildings, remain below. Quite modest transmitter power can therefore bridge vast distances. The primary radio transmitters on the Voyager probes have a strength of only 36 W and communicate with Earth from the edge of the solar system. In comparison to a walkie-talkie on Earth with its 5 W power, this is quite remarkable!

Some obstacles, however, do remain for radio transmissions in space. Planets and moons block radio waves completely, and they have the habit of filling a large section of the field of view when in orbit at low altitude. There are multiple options how to retain radio connection with a spacecraft orbiting a celestial body. The first is a suitable network of ground stations, so that

---

[3]The 5G standard is specifically designed to allow for handheld devices, cell towers, satellites and spacecraft to communicate with one another, so this might be changing at some point.

a line of sight connection to the craft can be upheld at all times. This is a realistic option only on Earth though, as other planets and moons would need a number of receivers on their surface for this purpose first. The other option is to build a relay satellite system, which links radio connections through further vantage points until eventually contacting a ground station. The American Tracking and Data Relay Satellite System (TDRSS) and the European Data Relay Satellites (EDRS) are two such systems able to bounce radio signals from one of their satellites to the next and thus allowing radio connection beyond line of sight.

There is currently only one satellite (the Chinese Queqiao) providing uplink to the backside of the Moon, specifically to keep connection with the Yutu-2 moon rover. Before it was launched, any spacecraft flying to the back side of the Moon (or even landing there) had no possibility to communicate with Earth at all! This is the reason why it is still referred to as "the dark side of the Moon", even though the Sun is shining onto it half of the month. Around Mars, the space probes Mars Odyssey, Mars Reconnaissance Orbiter and ExoMars Trace Gas Orbiter form a provisional data relay network that the rovers on the surface are using to communicate with Earth.

The speed of light becomes a real limitation, even when communication links are unobstructed, due to the vast distances of spaceflight. As nothing in the universe can move faster than 300,000 km/s, a bigger and bigger delay is incurred when trying to communicate with targets further away. In communications between the Earth and the Moon, this delay already becomes noticeable, with about a second of round-trip time. Between Earth and Mars, the one-way time lag depends on the relative positions of the planets on their orbit and can range between 3 min on closest approach and half an hour at largest separation. This makes direct back-and-forth conversations impossible. Instead, communication will have to occur asynchronously, through recorded voice, video or text messages. The furthest human-made object, the Voyager 1 space probe is more than 22 billion kilometres away from the Earth, so radio signals take over 20 h to reach it!

Data Transfers

No matter whether a flight to space departs for touristic, commercial, political interests or scientific reasons (the latter one being the dominant reason by far), it will produce plenty of data. This data can be photos, videos, experiment results, Earth observations or medical measurements of the spacefarers themselves. Also the spacecraft itself tends to generate large amounts of telemetry parameters to be analysed in ground control.

For some of these data, like photos, it is completely acceptable just to be stored on a hard disk and brought back to Earth after the expedition. Other data needs to be downlinked in real-time or transmitted on request. Since downlink speeds (and, on small satellites, also storage space) are limited, it is important not to make a mess of the data. In NASA terminology, this is called *On-board Data Handling*, and it involves a comprehensive specification of data formats, which the multitude of computer systems on board and on the ground use to communicate.

In addition to data being downlinked to the ground, there is a need to keep a data connection in upwards direction. In crewed spacecraft, somebody tends to be around to push buttons and to communicate over the radio. But who keeps a spacestation flying while the crew is asleep or currently visiting another planet? How do uncrewed vessels, like satellites, get their manoeuvring commands? Most functions of the craft should be controllable from afar. This leads to the idea of the *telecommand*.

A telecommand remotely instructs a computer to perform some function, for example the start of a data transfer, ignition of a thruster or increasing the volume of the life support system. Such a command can be executed immediately, but it can also be time-tagged, meaning that the command should only be enacted at a given time. Likewise, the command can require a certain location or velocity, for example, to take a photo of a specific point of Earth's surface. Bulk data transfers back to Earth are handled correspondingly, by starting the transfers using time-tagged telecommands whenever the spacecraft flies over a ground station. This is especially true if the ground station only has a single frequency to communicate with the satellite, and therefore cannot send and receive data packages at the same time.

It is important to be able to reset a spacecraft remotely using a telecommand, so it does not get stuck in a state where control and manoeuvres would be impossible. The radio subsystem is carefully designed so that no dead-end situations can occur, in which further commands get ignored (like switching off the radio or disconnecting all power sources at the same time).

For low orbit missions, telecommand scheduling can be avoided if permanent radio connection is available, but that comes with the cost of a satellite relay network or multiple ground stations scattered around the Earth—which remain the domain of large space agencies. In commercial utilisation of space, as well as in travels beyond Earth orbit, telecommands continue to be the trick of the trade to remotely control spacecraft.

Computers and Storage Media

One of the computers in a spacecraft is the Guidance and Navigation Computer (GNC), with the responsibility to determine the trajectory and to ignite thrusters and engines. It is usually joined by one or many payload computers, which have a diverse set of functions. They can supervise scientific experiments while gathering and processing resulting data, be in charge of life support and environment monitoring or keep inventory of supplies on board.

These computers are optimised to operate in space conditions. Cosmic radiation causes increased rates of computing mistakes, randomly fluctuating transistors and flipping of bit values in memory leading to garbled data, which the system needs to handle without information loss. In many cases the computers are designed to be redundant, meaning that every operation and calculation is performed independently by multiple computing units. The results are compared and errors rejected before any action is taken. The increased radiation doses compared to Earth also have an impact on the choice of storage media: Magnetic drums and tapes as well as magnetic bubble memory, all of which have long been replaced by flash memory in consumer products on Earth, are still very important in space. Magnetic storage is much less susceptible to radiation damage compared to semiconductor transistors (while a transistor easily gains or loses charge, a magnet does not swap its north and south poles simply because a radiation particle hits it).

Magnetic disks, such as hard disks, have to be modified for weightlessness for a different reason: Due to their shock sensitivity, hard disks on Earth carry acceleration sensors that automatically move the reading mechanics to a safe position when the system detects that it is in free fall. In zero gravity, this sensor has to be disabled in software, as it would otherwise permanently consider itself to be falling and never allow read or write access to the disk. (In addition, hard disks require a minimum gas pressure around them to keep their read/write heads suspended above the disk surface. They will not work in a vacuum and are only operated in pressurised spacecraft.)

Contrary to the prevalence of magnetic storage media, the Mars rovers Spirit, Opportunity and Curiosity indeed utilise flash memory in their onboard computers. Even though they suffered plenty of radiation damage and data loss on their travel to Mars, their software worked fine, as it was designed to withstand these conditions. Instead of storing data in memory only once and trusting that it stays there, everything is written to memory multiple times. Data is also checked for radiation corruption any time it is read back. With this method, the Mars rovers' computers survived throughout their entire mission duration.

For endeavouring spacefarers who would like to play with a spacecraft computer before starting their space adventure: There are at least two spacecraft computer systems, which are available for download and can be run on a normal PC. Specifically, these are the *Apollo Guidance Computer* (AGC) and the *Neptun-M* used in Soyuz Rockets.

- The complete source code of the Apollo computer system (the user interface is shown in Fig. 2.28a) was found in the archives of NASA, printed on stacks of paper. A large group of volunteers typed it back into digital form, and the virtual Apollo Guidance Computer project (vAGC) developed an emulator that allows running the original software on a regular PC. Figure 2.29 contains a link to the project.
- The Neptun-M computer of the Soyuz programme (since the year 2002) is in principle compatible with a regular PC and runs a DOS operating system. Figure 2.28b shows a snapshot of its software, and Fig. 3.6 contains a pair of these computers operating inside the cockpit of a Soyuz capsule. The Zvezda module of the International Space Station contains the same computer, which is responsible for controlling attitude and manoeuvres of the entire station. A version of this software for spacefarer training purposes (showing mostly fictitious data) can be downloaded from the link given in Fig. 2.30. But note that both the website and the software are only available in Russian!

a)   b)

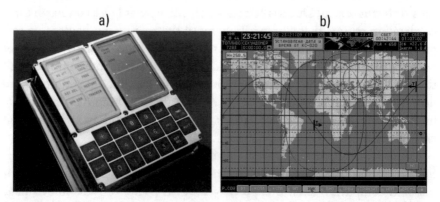

Fig. 2.28 Navigation computers have evolved in the way they display their information. (a) The user interface of the Apollo Guidance Computer only had a small numerical display (Image: NASA). (b) The Soyuz computer software shows a map view with the spacecraft ground track and the Russian radio coverage areas of the ground stations (small misshapen circles over Russia) and a geostationary relay satellite (large red oval) (compare the displays in Fig. 3.6)

Fig. 2.29 The virtual Apollo Guidance Computer project provides the full source code of the Apollo spacecraft computer, as well as an emulator to use them on a normal PC. https://www.ibiblio.org/apollo/

Fig. 2.30 The URL http://astronaut.ru/bookcase/prog.htm leads to a somewhat plain website, completely in Russian. A cosmonaut training version of the Soyuz computer software is available for download

Anecdote

When the Apollo navigation computer was developed in the 1960s, it was the first time when many tasks were handled with a computer. A number of new programming concepts and ideas were invented by the team led by computer scientist and mathematician Margaret Hamilton. Some of these concepts and ideas were forgotten and only rediscovered many decades later in other pieces of software. One such example is the introduction of dynamic memory management, which allowed multiple processes to run at the same time on one machine. Along with it, the problem of a "memory leak" was discovered:

Since computers were still large and heavy during the Apollo era, the decision was made to use only a single computer to perform all relevant control tasks, such as regulating the engine throttle, measuring the altitude with a radar system, supervising pressures and voltages and keeping instrument displays up to date. All of these functions had their own programmes, between which the computer rapidly switched back and forth, based on their priorities in the moment. The whole setup had been designed and tested so that it could run throughout the entire lunar landing without the programmes interfering with each other.

(continued)

> When the first moon landing actually happened, things did not quite turn out as planned. In the flight of Apollo 11, the astronauts Neil Armstrong and Buzz Aldrin forgot to switch off the rendezvous radar of the lander (which was only going to be required for redocking with the orbiting command module later on), so the computer had to run one additional programme than originally envisioned, for reading and displaying the radar distance. Since this was considered a low priority programme, other processes were continuously interrupting it, making it unable to process the incoming data completely. It started to fill up free memory, until none was available anymore... whereupon the computer crashed, showing only "Error 1201". It restarted automatically and triggered a loud warning sound inside the spacecraft.
>
> The astronauts were understandably very concerned, as they were in free fall towards the surface of the Moon at the time. An engineer in the ground control station in Houston thankfully understood the cause of the problem and decided that the error would not endanger the landing process. Apollo 11 eventually landed safely on the Moon, with loud error sounds and a computer that was crashing every couple of seconds.

The less critical an electronic device is for mission success, the more ordinary its design can be. Nowadays, more computers with mundane applications are present in addition to those with important control functions. Less sensitive systems serve purposes like daily schedules, logbooks, entertainment, etc., and they tend to be quite similar to machines on the ground, with only some minor modifications still required for error correction of memory and flash storages. Digital cameras are examples of electronic devices that are left unmodified for spacecraft (with a somewhat increased image noise due to scintillation from cosmic rays). Entertainment electronics, such as CD players, laptops and tablets, have been flown to space as such. Nowadays, the International Space Station contains a wireless LAN to connect with the crews' personal electronic items!

2.7   Further Equipment and Considerations

Sitting, Standing and Eating: Furniture in Space

Tables, chairs and beds are not in principle necessary in weightlessness. Simply drifting freely in the cabin is just as comfortable, or even more so, than sitting or lying down is on Earth. Yet, there are some helpful constructions that have proven to facilitate the stay in space.

Floating aimlessly around the spacecraft, as enjoyable as it might be, is quite impractical for many tasks. Oftentimes, it is better to be fastened in place, to keep everything, including the person itself, in the work area. Also when performing mechanical tasks such as bending or drilling, being anchored helps to exert physical force onto the object instead of starting to spin oneself. It is therefore recommended to scatter handlebars and rope loops around the cabin walls, to hold onto with the hands or to hook into with the feet.

Any available flat surface can be used as a working area in zero gravity, without a reason for it to be table-shaped, but a table fulfils an important social function: Gathering the entire crew around a warm meal and "sitting across" one another helps with the group's social cohesion. The table does not need to be particularly flat, level or solid for this—after all, no weight is actually resting on it. It can as well be a foldable or slideable board. The dinner table of the Skylab spacestation (Fig. 2.31) was built as a central, collapsible column. Food trays were mounted onto it, resulting in a still pretty archetypical table shape, which has since fallen out of use.

Traditional seats or chairs also are without any actual function in zero gravity. The concept of "sitting" in itself makes little sense in space: When trying to sit somewhere without gravity, one ends up bouncing off or simply floating a small distance above the surface of the seat. This allows weight and storage space to be conserved because sitting furniture does not even need to be

Fig. 2.31 The dinner table of the Skylab spacestation's ground training facility. A central column can be extended from the floor and food trays can be fixed to the side (Image: NASA)

considered. In the Apollo lunar lander, for example, no seats were provided for the astronauts while they descended towards the surface of the Moon, as they could comfortably pilot the craft standing up and holding on to handlebars (the Moon's surface gravity is only 1/6th that of Earth).

Only when returning to Earth and entering into the atmosphere does comfortable sitting gain importance again, as both reentry and landing involve shaking and strong G-forces acting on the spacecraft. The spacefarers, having spent an extended period of time in weightlessness where their muscles and bones have grown unaccustomed to loads, should be cushioned as well as possible to lessen and equally distribute the strain encountered on the way back. These chairs can be constructed as anatomically fitting clamshells (as is the case in a Soyuz capsule) or designs similar to sports car seats. They can also be mounted with shock absorbers or crush cores (see Sect. 2.2) for extra buffering.

Waste Removal and Space Debris

Supplying a spacecraft or a spacestation with necessities and consumables is a veritable logistical problem per se, but it automatically leads to a second one: What to do with the waste that accumulates? Opportunistically, one would hope that it can simply be tossed out of an airlock into space, and it would drift away, never to be seen again. Unfortunately, orbital mechanics has a thing or two to say in this matter (see Chap. 3) and turns every piece of trash into a potentially dangerous space debris projectile. If the trajectory is not carefully considered, any item thrown out of an orbiting spacecraft will collide with it later on within the same orbit! Hence, unthoughtfully throwing trash out of an airlock is not a good idea.

Neither can the rubbish be stored inside the craft for all eternity (very short flights do not need to deal with this issue). For spacestations, the most common and easiest way of rubbish disposal is fulfilled by the same transport craft that brought supplies up into orbit in the first place. Stuffed with trash before undocking, most of them burn up in the atmosphere, turning to ashes and scattering into the wind as dust (SpaceX's Dragon transporter is the exception and lands in the Atlantic Ocean. Any trash stored in it is ordinarily disposed of on Earth).

For expeditions further into the solar system, for example, towards Mars, this concept cannot be continued as such, as no resupply vessels will visit on the way. Trash capsules, which can be as simple as a stable plastic bag with a small thruster unit attached, might be required. The waste can be directed away from the vessel and into a collision course with the target body (alternatively, it can be flung out into space to orbit the Sun forever). As long as interplanetary flights are rare occurrences, there is no need to worry about polluting the solar system—Earth gets hit by 15,000–78,000 tonnes of meteoroids every year anyway. Should space travel become an everyday commodity in the future, better waste handling methods need to be available!

One possibility to reduce the waste volume significantly is by either decomposing or incinerating it on board. Both these processes release gases which can be vented into space or even be employed for manoeuvring, either way dissipating into space and being blown away by the solar wind. The remaining ashes would cause a much reduced waste problem. NASA's *Heat Melt Compactor* project actually suggests to form the residue into compact tiles and use them for radiation shielding. Pyrolysis and decomposition processes in zero gravity behave quite a bit differently than on Earth, so this is an ongoing field of research.

Spacecraft themselves, especially uncrewed satellites in low Earth orbit, are causing an increasing trash problem. Since the beginning of the Space Age, thousands of satellites, probes, rocket stages, loose screws and pieces of debris from anti-satellite weapon tests have thus started to accumulate around the Earth. Many of them were left to fly until they stopped functioning (typically by either running out of propellant or because their electronics stop working due to radiation exposure), and keep circling on their orbits. As soon as any two items collide, they tend to break up into smaller pieces, which further increases subsequent collision probabilities. Some debris clouds were created quite deliberately, such as the *Westford needles* experiment, in which 480 million tiny copper needles were dispersed around the earth to act as radar reflectors. While solar wind pressure has pushed most of these flimsy pins from orbit, some 40 clumps of these needles are still around, over 60 years later. To prevent the problem from getting out of control, space missions nowadays need to ensure that the formation of new space debris is minimised. This can either be achieved by placing objects into orbits that naturally decay through atmospheric drag after their intended mission duration, alternatively by boosting them to an equatorial *graveyard orbit* high above the geostationary level, where they orbit in the same direction and pose no danger to crewed spaceflight or by manoeuvring them to Earth before propellant runs out. When the latter option is chosen, reentry is usually aimed towards the middle of the Pacific Ocean (at a location called *Point Nemo*), where they are least likely to hit any inhabited area.

Anecdote

While performing a spacewalk for repairs on the International Space Station in November 2008, the American astronaut Heidemarie Stefanyshyn-Piper forgot to clip a tool bag onto a holding rail on the station's surface. The bag slowly started to drift away, and the astronaut only realised this when it was out of reach.

The bag continued to fly by itself on an orbit in proximity to the station and came close to it again 45 min later, but the astronaut was unable to climb the outsides of the station fast enough to catch it (moving in a spacesuit is exhausting, see Sect. 4.5).

The lost tool bag circled Earth on its own for a couple of months while slowly sailing away from the ISS. It was photographed from the ground by amateur astronomers and used as a test object for space debris tracking. It eventually got slowed down by the tenuous atmospheric friction, reentered in August 2009 and burned up.

2.8   The Launch Site

Appropriate ground infrastructure for a rocket launch site goes beyond some fuel pumps and a ladder to climb up into the spacecraft. Since rockets' exhaust is a hot and powerful stream of gas, the launch site needs to be able to sustain the onslaught, if it is intended to be used more than once. NASA traditionally builds launch pads out of heat-resistant steel and concrete (Fig. 2.32). The pad has a hole directly under the rocket engines which leads to a *flame trench* that directs the rocket exhausts sideways, so that they leave the pad without melting its structure. A monumental arrangement of sprinklers floods the entire steel construction and trench with water just before launch, cooling their surfaces and producing copious amounts of vapour once the engines ignite. The Russian approach to launch sites is slightly different, as in their design the rocket is hanging over a gigantic pit in the ground and omits a water cooling system. Once the rocket engines have successfully ignited, the holding mechanism releases the rocket to swoop spacewards. Hence, American rocket launches are accompanied by huge clouds of white water vapour engulfing the launch structure, whereas Russian launches show only relatively small amounts of soil-coloured dust and swirls of heated air.

Similarly, big differences exist in the ways that rockets are assembled and brought to the launch site. Soyuz rockets and the Falcon rocket family of SpaceX are constructed while lying down horizontally, delivered to the launch site on a train (Fig. 3.1) or trailer and swivelled into vertical position with the help of an erector arm. Both NASA and the China National Space

Fig. 2.32 Launch Pad 39A at the Kennedy Space Center. A total of 12 Saturn V and 81 Space Shuttle flights launched from here. It is currently used by the Falcon 9 and Falcon heavy rockets. The concrete ramp leading up to the launch platform, with its 250 m length and a slope of 5% allows rockets to be delivered upright on a crawler. The flame trench under the platform (not visible in this photo) is 13 m deep, 18 m wide and 137 m long. The water tower for the cooling system is 88 m high. The pad has undergone modifications to accommodate vehicles by SpaceX (Image: B. Ganse)

Administration (CNSA) prefer to directly piece their rockets together in a vertical launch configuration, which requires extremely tall buildings close to the launch site, such as the *Vehicle Assembly Building* (VAB) in the Kennedy Space Center. In these cases, the entire mobile launch platform is moved to the launch site from the assembly building, for example by the *crawler* shown in Fig. 2.33.

Fig. 2.33 A *crawler* under a mobile launch platform at the Kennedy Space Center. The crawler is able to carry the entire platform including a rocket to the launch site via the crawlerway (Image: B. Ganse)

Another approach has been chosen by SpaceX to build the Starship spacecraft family. Due to their corrosion-resistant (stainless steel) hulls, Starships can be constructed out in the open without the need for an assembly building. They are meant to launch directly from a concrete pad at their building site. Even though the concrete pad has to be repaired after every launch, the end result is a net reduction in cost and complexity.

References

Clark, J. D. (1972). *Ignition! An informal history of liquid rocket propellants*. Rutgers University Press. ISBN 978-0-8135-0725-1.

Janhunen, P. (2004). Electric sail for spacecraft propulsion. *Journal of Propulsion and Power, 20*(4), 763–764. https://doi.org/10.2514/1.8580

3

How to Fly a Spacecraft

Contents

Endeavouring spacefarers that have read the previous chapter carefully, possess some mechanical skills and a suitable workshop, might now have their own spacecraft available or underway. So it is high time to learn something about steering, trajectory planning and navigation in space! Flying a spaceship is not much more complicated than steering a car or a ship—it is just determined by different environmental conditions and physical laws that have to be understood first.

3.1   Preparation, Countdown and Launch

Before a rocket can launch into space, plenty of groundwork is necessary to make the flight go both smoothly and safely. While the details of launch preparation steps differ from one spacecraft and launch site to another, a rough overview of the elementary phases will be given here, from the arrival at the launch site (Fig. 3.1) until the blastoff into space.

© Springer-Verlag GmbH Germany, part of Springer Nature 2020
B. Ganse, U. Ganse, *The Spacefarer's Handbook*, Springer Praxis Books,
https://doi.org/10.1007/978-3-662-61702-1_3

Fig. 3.1 Russian Soyuz rockets are delivered to the launch pad by train and only lifted into vertical orientation once arrived (Image: NASA)

Countdown

The countdown is probably the most well-recognised phase of a space mission. It is the part of a crewed mission that media most often portray, to the point that it has become the symbol of the conquest of space. The invention of the countdown is actually older than spaceflight itself: It was first shown in the 1927 movie "Woman in the Moon" by Fritz Lang, where a rocket launch was presented as dramatically as possible.

The location where the countdown is conducted is the launch control room. Here, all relevant information is gathered together, before a decision is made to issue the clearance for launch. This data includes, for example, system diagnostics from the spacecraft, medical readings of the crew, wind strengths, weather predictions and the *launch window*, a time range where the launch to the desired target orbit is possible (see Sect. 3.3). At NASA, this is done in the *Launch Control Center* (LCC) at the Kennedy Space Center (Fig. 3.2). All activities before the actual launch are performed in a strictly structured timetable, the countdown.

While media typically only show the last 10 s, the entire procedure is significantly longer and more complicated. In current rocket launches, the

Fig. 3.2 A typical workstation in the *Launch Control Center* at the Kennedy Space Center as it was during the Space Shuttle era. From here, launches were monitored and controlled. As soon as the crewed spacecraft have launched, control is handed off to the *Mission Control Center* (MCC) in Houston, Texas (Image: B. Ganse)

count is normally started multiple days before the intended liftoff time. Furthermore, different times are used: The "L-minus" steadily counts the actual amount of hours, minutes and seconds until launch time without any stops, but this time is never really used in any official manner (in case of a launch delay, the time is simply rewound to countdown towards the new launch time). The launch processes' work schedules are written down in terms of the "T-minus". While this number also counts down towards the moment of liftoff, it can occasionally be put on hold to allow for delayed work to catch up. In Space Shuttle launches, for example, the countdown clock was stopped at "T minus 10 minutes" to have a scheduled, 30 min hold time during which all incomplete tasks could be finished. In some cases, the count can even be moved forward if everything is progressing faster than expected. The launch director ensures that all systems are "go for launch" before restarting the clock and entering the terminal count phase. Here finally, the L- and T-time are identical and continuously count down towards the moment of liftoff. In the end, it is a tradition for the launch director to announce: "Ten seconds to liftoff. T minus nine, eight, seven, six, five, four, three, two, one. Liftoff!" (or:

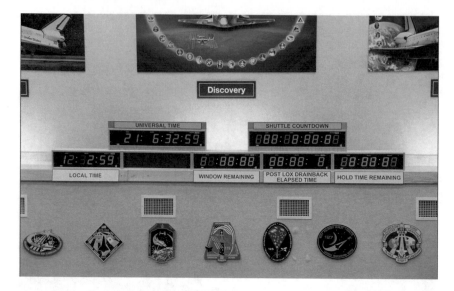

Fig. 3.3 Time displays in the *Launch Control Center* at the Kennedy Space Center. *Universal time* = time in days, hours, minutes and seconds since the start of the year. *Shuttle countdown* = time until launch (T-minus). *Local time* = time in Florida. *Window remaining* = time until the launch window closes. *Post lox drainback elapsed time* = remaining time the spacecraft can stand on the launch pad while filled with liquid oxygen. *Hold time remaining* = remaining time until the T-countdown continues (Image: B. Ganse)

Ignition/Décollage/Poyekhali, depending on the launch country), and then to finally let out a sigh of relief (Fig. 3.3).

After launch, the mission clock continues counting seconds, minutes and hours. Now, it is no longer counting down numbers until launch, but counting up engine burn time. After engine cutoff, once the spacecraft has reached orbit, the same time usually gets referred to as mission elapsed time (MET).

Anecdote

What does the crew do in the evening before the launch? In Baikonur, Kazakhstan, the Soyuz crews traditionally watch the 1970 movie "The white Sun of the Desert", which is considered a cult classic of Soviet movie making. Originally it was shown to the cosmonauts as an example of good camera work to improve their filming skills on orbit. It is considered to bring good luck for the flight, ever since the catastrophic flight of Soyuz 11 (see Sect. 5.6) had skipped the movie.

Dramatic Effects Around Liftoff

Photos and videos of rockets already before launch often show them hissing out steam, both from the engines but also from the upper ends of the fuel tanks. This results from the choice of rocket propellants, such as liquid oxygen and hydrogen, which are stored in the tanks at cryogenic temperatures. Since the tanks are made from astonishingly thin metal sheets to be as light as possible, they are not particularly good at insulating heat and the liquefied gases evaporate. The propellants are literally boiling off, as the ambient air has a temperature hundreds of degrees too high for them to stay liquid. To prevent the tanks from bursting, excess gas is vented from the top ends of the tanks through an overpressure valve. Humidity in the air condenses on contact with these cold gases, making them visible as steam. A couple of seconds before engine ignition, rocket engines with actively cooled nozzles (see Sect. 2.3) start pumping their coolant, such as liquid oxygen, through the engine for the *pre-ignition chilldown*, leading to additional vapour puffs at the bottom end of the rocket. In some cases, sparklers on the launch pad are activated to burn away the excess oxygen before engine ignition.

The second side effect of putting cryogenic liquefied gases into badly insulated tanks is the formation of ice on the outside of the rocket, especially in launch sites in humid climates, such as in Florida and French Guyana. Strong vibrations during liftoff tend to shake chunks of ice off the rocket that are clearly visible in close-up photographs of Apollo launches.

The Space Shuttles were an exception to this. Even though they were fuelled with cryogenic oxygen and hydrogen, they did not have ice forming on the outside of their main tank. The engineers were concerned, that ice falling off the tank could damage the ceramic heat protection tiles on the orbiter (in retrospect, quite correctly so, as the destruction of the Shuttle Columbia showed, see Sect. 2.2). The main fuel tank was therefore padded in insulating foam, which gave it its characteristic orange colour.

Liftoff

During the last seconds of the countdown sequence, the engines are finally started up and sensors and computers check that all systems are functioning correctly. It is not unusual for a countdown to be aborted only 3 s before liftoff, for example, if a pressure reading inside a combustion chamber was a little bit off!

Fig. 3.4 Slow-motion video of the processes taking place on the launch pad at and around the moment of liftoff of a Saturn V rocket. https://www.youtube.com/watch?v=DKtVpvzUF1Y

When the count finally reaches zero, any hold down clamps of the launch pad are released, umbilical cables and fuel lines are detached and the rocket lifts off. A narrated slow-motion video of a Saturn V launch, linked in the QR-code of Fig. 3.4, shows the complex events on the launch platform at liftoff.

After liftoff, when the rocket has left the launch pad and cleared any towers or structures around, ascent through the atmosphere is the next stage of a spaceflight. The first task is now to gain altitude as quickly as possible. Before starting to pick up any kind of orbital speed, the dense low altitude atmospheric layers with strong friction and high pressure need to be left behind. Even when launching from a moon without an atmosphere, a minimum upwards bounce is needed to avoid geographical obstacles like mountains or crater rims. One should then, relatively quickly, start nudging the flight path in the desired downrange direction (more about this in Sect. 3.3) so that the rocket slowly inclines.

This leads to a gradual transition from a vertical to a horizontal orientation, the *gravity turn*. In an atmosphere, the gravity turn has the useful characteristic that the rocket always flies with its tip pointed straight into the wind that results from its own upwards motion, so aerodynamic forces are only acting along the long axis. Thus the craft never has to withstand any strong sideways wind effects.

Launch and ascent are the most dangerous flight phases, as the massive engines of the first stage produce tremendous amounts of thrust and any malfunction can lead to an enormous explosion. In order to nonetheless ensure the survival of the crew, basically all crewed spacecraft contain some kind of launch escape system, which is able to bring the crew away from the rocket and into safety in case of emergency. In most cases, this is implemented as a small solid rocket tower at the tip of the vehicle that is able to pull the

Fig. 3.5 The launch escape system of the Orion capsule is a small solid rocket sitting at the top of the rocket. In case of a launch emergency, it pulls the capsule away from the rest of the rocket with a short but strong acceleration. Afterwards, the parachutes open and the capsule lands safely in a couple of kilometres distance from the launch site. Similar systems exist in nearly all crewed spacecraft (Image: B. Ganse)

capsule upwards and to the side, away from the lower stages (see Fig. 3.5). SpaceX's Dragon spacecraft carries powerful thrusters for propulsive landing, which double as the launch abort system and make a separate escape tower unnecessary. In the Gemini capsules, rocket-powered ejection seats for the crew were provided instead.

Launch escape systems are in no way comfortable to use, as they accelerate violently (up to 14 G) for a few seconds in order to get out of the danger zone quickly. They are used extremely rarely: The Soyuz launch escape system was triggered only twice. In September 1983, during the Soyuz T-10a mission, the rocket started leaking kerosene a couple of seconds before engine ignition, caught fire and caused a "rapid unplanned disassembly" (an engineer euphemism for a detonation). Both cosmonauts remained unharmed, thanks to the rescue mechanism. The second case of a launch escape firing occurred in October 2018, when the booster separation of Soyuz MS-10 failed and the central rocket stage was damaged. The launch escape system again brought the crew safely to the ground.

The Space Shuttle did not come with an automatic launch rescue system of any kind—when its solid rocket boosters were burning, no launch escape was possible, which had fateful results in the Challenger catastrophe, where the first stage tanks exploded shortly after launch. There was no way to save the crew of seven. After this incident, a rudimentary launch escape mechanism was added, which involved opening the door and jumping out with a parachute, see Fig. 4.20.

One very uncomfortable job in the American and European space programmes is the Range Safety Officer (RSO). This person has the responsibility to make sure that the launching rocket does not endanger the general public. To do so, the RSO has a console, from which they can trigger explosives on the sides of the fuel tanks to blow up the rocket in flight! More than 30 launches of uncrewed spacecraft have ended with such a flight termination. This self-destruct system is armed even in crewed spaceflight, with the difference that the explosion is delayed to give the crew escape system sufficient time to act (3 s in the case of Mercury, Gemini and Apollo launches). The Russian and Chinese space programmes' range safety systems do not include explosives and instead shut down the engines when the rocket deviates from the intended flight path. At least in one case, this caused a Chinese rocket stage to impact and explode in a small town, leading to multiple fatalities.

Stage Separation

Ejecting engines and fuel tanks to reduce total spacecraft mass as soon as they are no longer required, is a common stratagem to make flights to orbit possible. Since the rocket nozzles' expansion ratio should fit the ambient conditions of the current flight phase (see Sect. 2.3), rockets are built of multiple stages, each of which ignites, burns for a predetermined amount of time and is then jettisoned. In the case of the first stage, the engine cutoff and jettison happens in dozens of kilometres of altitude, where the remaining air pressure still causes some leftover resistance. As a result, the *stage separation* process needs to balance several effects with each other and forms a compromise (for further stages that are detached at much higher altitudes, the same considerations apply, albeit with much reduced atmospheric effects):

- Switching off the previous stages' engines prior to separation means that no thrust is generated for a moment, while air friction continues to impose drag on the craft. Hence, the spacecraft loses velocity during stage separation. To minimise this effect, separation should be as quick as possible.

- A rocket engine does not simply instantaneously turn off. Leftover propellants in the fuel lines and combustion chamber continue to produce a small amount of thrust for a few seconds. When separating rocket stages, it should be ensured that a just-detached first stage does not get pushed right back into the upper part of the craft and damage it (as happened in one of the first launches of a Falcon rocket).

- Once separation is complete, the engines of the next stage ignite and blow their hot exhaust downstream. If the waiting time between separation and ignition is not long enough and the exhaust gases hit the detached stage, this can cause problems or even an explosion of the jettisoned stage! The detached stage can either be moved out of the way sideways or, like the Soyuz rocket, carry a heat shield on its upper end.

- Between engine shutoff of the previous stage and ignition of the next one, the spacecraft is in free fall for a short span of time, meaning that the entire contents suddenly behave like in zero gravity. This includes liquid fuel tanks, used for the next stages' rocket engines. As no forces are holding the liquids down anymore, they are now rapidly distributing throughout the entire tank volume, like the contents of a water bottle thrown into the air. This situation, called *fuel sloshing*, causes a similar problem to the fuel tanks of engines used in space (see Sect. 2.3), where gas bubbles might be ingested into the engines and cause problems (asymmetric thrust or even an explosion). During ascent of a spacecraft, there are some additional ways this situation can be handled:

  - One option is to have a set of small solid rocket engines, called *ullage thrusters* mounted on the upper stage, simply to give it a short burst of acceleration, so that fuel again collects in the bottom of the tanks and the engines ignite reliably. The rockets of the Apollo programme were routinely using this design, while the SpaceX vehicles are achieving the same effect using cold gas thrusters.

  - The alternative solution applied by Soyuz rockets is to ignite the second stage engine already before stage separation, with a low throttle setting (this is a further reason why a heat shield is required at the top of the first stage). The free fall phase of the rocket is completely circumvented this way.

  - While capillary forces can "hold the fuel in place" in the fuel tanks to prevent it from sloshing away from the engine inlet, this only works properly for small engines and accelerations. Due to the size and mass of the vehicles employed in crewed spaceflight, other methods are typically preferred over this one.

Technically the stage separation tends to be done by compression springs. In their pre-separation state, the springs are kept compressed by steel wires that are equipped with tiny explosive charges. When the stages separate, the charges are detonated, cutting the wires and causing the springs to fling the stages away from each other. Another option is a pneumatic system that employs gas pressure, or even residual pressure through a valve at the top of the fuel tanks to push the stages apart.

Once upper strata of the atmosphere have been reached, where aerodynamic forces like resistance and lift are negligible, the previous requirement for the spacecraft to face in the direction of travel is no longer valid. Depending on the intended trajectory, another orientation might be better for thrust manoeuvres. Furthermore, any kinds of protective shells, launch escape systems, control surfaces or other aerodynamic helpers can now be ejected to reduce mass (except if the spacecraft is meant to reuse them).

The engines' job, however, is far from done when the atmosphere has been left behind! As Sect. 3.3 will explain in more detail, a stable orbit does not only require the spacecraft to be flung high enough, but also to progress in a curved flight path that results in a tremendous velocity around the planet. Continuing to accelerate towards the correct orbital velocity takes the lion's share of the engine run time. The amount of energy to lift a kilogram of mass to the height of the International Space Station is actually lower than the energy needed to accelerate it to orbital velocity!

3.2   Steering in Space

Cars, ships, airplanes, bicycles and horses—all of these earthbound means of transport known from everyday life have a conceptually similar way to move. They have a clearly defined "front" and "back" direction and typically move head first. To reach a desired destination on Earth, all one needs to do is to point the means of transport towards the target, pick up speed and wait until arrival.

This strategy does not work with spacecraft, even though science fiction would want to believe in it. To get from one point to the other, the first thing one needs to do is to stop thinking about straight lines. Instead, the interplay between thrust and gravity has to be considered, which primarily manifests itself in curves such as ellipses and hyperbolas. This chapter claims to make said interplay understandable and to enable even the layperson to fly a spacecraft.

First of all, the most important rule is:

Stay calm and keep breathing.

When flying a spacecraft, there is always enough time to thoroughly ponder what actions to take. Single manoeuvres are usually more than half an hour apart, leaving enough time for consideration.

Joysticks, Control Handles and Thrusters

When looking at the cockpits of spacecraft used in human spaceflight (for example, Fig. 3.6) two things are quickly noticeable:

1. They are filled with ridiculous amounts of switches and buttons (or nowadays, touch screens).
2. They are steered using two joysticks (or similar grabbable controls, one for each hand)

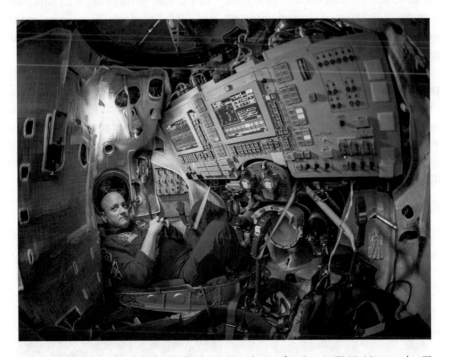

Fig. 3.6 NASA astronaut Scott Kelly in the cockpit of a Soyuz TMA-M capsule. The round object next to the astronaut's knee is a viewing screen, which allows the pilot to look into the direction of the spacecraft's docking port for rendezvous and manoeuvring. The two flight control joysticks are located on both sides of it (Image: NASA)

The first observation is simply owed to the complexity of the systems present in a spacecraft (compare Chap. 2) and is especially present in designs from the '60s and '70s, when automation of systems was still limited and manual control of every subsystem was the norm.

The second point raises more curiosity: Spacecraft cockpits come with two hand controllers, one for the left and one for the right hand. They may not always have a joystick shape, sometimes built in a form that resembles a valve handle or more of a solid block style. Why is this so? One reason, of course, is that people tend to have two hands. Nonetheless, spacecraft have been built by completely different nations over multiple decades and yet the control elements are basically identical! To apprehend this, it is important to first consider what it precisely means to steer in space.

As elucidated in Sect. 2.3, rocket engines are currently the only practical method to propel a space vessel. In addition to the big main engines, which are all installed to face the same direction, a multitude of smaller rocket nozzles are scattered around the hull of the ship, called *control thrusters*. The complete system, consisting of control thrusters, fuel feed and electronics is referred to as the *Reaction Control System* (RCS). Figure 3.7 shows examples of thruster groups on an Apollo service module and the nose of a Space Shuttle.

At first sight, the thrusters seem to be arbitrarily scattered around any possible corner of the vehicle, each facing into a different direction. Their arrangement does, however, follow a simple logic (demonstrated in Fig. 3.8 in a simplified way). By placing the thrusters as far from the mass centre of the vessel as possible, they can easily make it rotate when fired. A thruster at the nose of the vessel "pushes on the nose" while an aft thruster "pushes the stern around". The thrusters can be activated in two different ways:

a)   b)

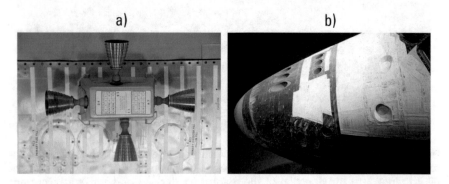

Fig. 3.7 Examples of control thruster arrangements: (a) One of the thruster quads on the Apollo service module. (b) The nose of a Space Shuttle orbiter. In both cases, thrusters are arranged in all possible perpendicular directions (Images: B. Ganse)

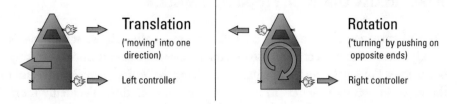

Fig. 3.8 The two primary modes, in which control thrusters are employed for linear motion ("translation") and for turning ("rotation") of a spacecraft

- When activating a rocket nozzle at the nose and stern on the same side, the ship is pushed by the recoil (compare Fig. 2.11) from both ends equally. The turning forces (moments of inertia) thus cancel out, and no overall rotation is created. The resulting sideways push is called a *translation*.
- If nose and stern thrusters are fired on opposite sides of the craft, their thrust acts as if bow and stern of the spaceship were pushed into different directions: Each of the thrusters tries to accelerate the craft, but is counteracted by the other one. No linear acceleration occurs, but the turning effect of both thrusters adds up leading to a *rotation*.

Exactly these two functions are provided by the two controllers in every spacecraft cockpit built so far. The right-hand joystick turns the spacecraft around its axes (usually it is built to be both tiltable and twistable, to steer all three axes of the ship) and the left-hand stick accelerates it in all directions of space (as it can be pushed and pulled as desired). When operating these controls, the same general principle as before applies: Stay calm! Wildly yanking them around will only lead the spacecraft into an uncontrolled tumbling motion.

Hint for Aspiring Spacefarers When trying to take control of a tumbling or spinning spacecraft, the first step should be to reach for the right-hand controller. It should be tilted against the spin direction to stop all rotation and align the craft with a well-defined direction in order to gain a comfortable understanding of where "up" and "down" is (defining these two in terms of an orbit is left to individual choice). Only after the spacecraft attitude is under control should one reach for the left stick and start performing navigation manoeuvres.

When to Use the Main Propulsion System

Control thrusters alone are not especially strong—after all, their purpose is mostly to turn the spacecraft and to execute precise movements during docking operations. For any larger manoeuvres, i.e. tilting the orbital plane or traversing the expanse between planets in a tolerable amount of time, decidedly more powerful engines are required. Basically every spacecraft is equipped with one or a cluster of vacuum rocket engines, which define its "forward" and "backward" direction.

The main propulsion system is primarily for manoeuvres which require significant changes in velocity—and in most cases, the engines are run at full throttle since it both makes their thrust more predictable and simplifies navigation processes (which tend to be based on the assumption that orbital changes happen instantly in one point, see Sect. 3.3). Navigating in space consists of a series of short engine firings, separated by long waiting times. In crewed spaceflight, each such navigational manoeuvre tends to be announced with a countdown of its own to make sure the crew is aware of the acceleration.

3.3   Orbital Navigation

As there is neither air nor any other medium to react against, a vessel experiences no friction in space. Without friction, everything that has been given a push continues to move indefinitely in a straight line. Based on this, it would seem that navigation in space simply consists of drawing a line between the starting point and the destination and following it strictly, as long as a sufficient supply of snacks for the way is on board. In free, empty space beyond the solar system, outside of the Milky Way, far away from any concentration of mass this actually would be the case. The current reality of human spaceflight however—and probably during the next couple of centuries—takes place inside the solar system, where flight paths are strongly affected by the gravity of the Sun, Earth and all other celestial bodies.

"Being in space" therefore means mostly the same as being on an orbit around a planet, a moon or the Sun. But what is an orbit anyway?

The Orbit

If one were to board a rocket, launch straight upwards and reach the altitude of the International Space Station, the stay in space would be very short: Earth's

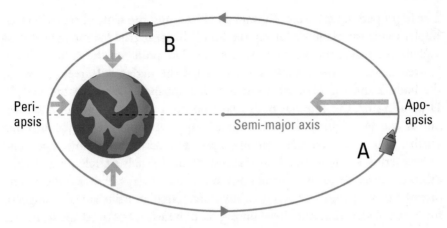

Fig. 3.9 A spacecraft in orbit around this Earth-like planet permanently "falls" around it. The ship at position A is moving around the planet fast enough to miss it while falling down, so by the time it reaches position B, gravity (orange arrows) is pulling it towards a completely different direction. Finally, it has fallen around the planet and returns to its starting position to continue on the same path. The semi-major axis, half the length of the longest diameter of the ellipse, determines the orbital time

gravity would simply pull the spacecraft back to the surface. It would fall like a rock for 400 km and reach the ground in a matter of minutes. Apparently, simply vaulting to a high enough altitude is not yet sufficient to get a spacecraft into a stable orbit. The decisive trick is to move around the planet at breakneck speed, in order to fall right past it!

Figure 3.9 illustrates this process: A spacecraft is flying around an Earth-like planet, with gravity (depicted by the orange arrows) acting on it. The craft is travelling through point A with a sufficient velocity to fall right past the planet. When it has reached point B, it has already passed a third of the way around the globe and gravity no longer pulls it into the same direction at all! In fact, the craft continues being pulled towards the centre of the planet all the time and returns to its initial point A. The same story repeats anew and it keeps missing the planet indefinitely. Just like riding an infinite roller coaster loop, the spacecraft gains more momentum as it approaches the periapsis and slows down when it climbs towards the highest point. The German astronomer Johannes Kepler discovered this phenomenon already in the year 1609 and formulated it as the *Kepler's first law*:

> Planets move on elliptical trajectories, where the Sun is located in one of the ellipse's focal points.

The flight path on an orbit is always an ellipse, and the same observation that Kepler made for planets orbiting the Sun is likewise true for moons orbiting a planet or spacecraft orbiting an asteroid. The point on the ellipse, that is located closest to the central body, is called the periapsis (Greek, "close to the body") and the furthest point is called apoapsis ("far from the body"). Depending on the celestial body that the orbit is revolving around, these points can get slightly different names: Apo- and perigee (when orbiting Earth), apo- and perilune (Moon), apo- and perihelion (Sun), apo- and pericytherion (Venus), etc. A special case of an ellipse is a circle. In a circular orbit, the distance to the central body is the same everywhere and there is no distinguished apoapsis or periapsis anymore. Circular orbits are advantageous for Earth observation satellites (among them weather, ocean temperature, ice cover, vegetation, pollution monitoring and emergency beacon tracking), as they maintain the same distance throughout the whole orbit.

One central idea in Kepler's first law that impacts orbital navigation is the fact that orbits are always closed curves. Specifically, if a spacecraft passes through a certain point in space on its orbit, it will pass through the same point again on the next and any subsequent rounds, if no other forces are acting on it beyond the gravitational pull of the central body. Weather satellites are directly making use of this feature: They are circling the Earth on orbits that lead them over the poles. While their trajectory always carries them through the same fixed path in space, Earth is slowly rotating beneath them with a period of 24 h, allowing them to scan a strip of Earth with their cameras and encounter every area on the surface twice a day. After completion of one 90 min revolution around the planet, the next overpass scans an area further westwards, as Earth has rotated by several degrees in the meantime. By circling over the poles, going from the south pole to the north pole and continuing southward on the other side of the planet, the satellites photograph each spot on the surface with both legs of the orbit: once while going northwards and again on a southward part of their tour, half a day later. The image strips are then combined on the ground to create a weather map of the entire Earth.

For any sort of manoeuvre along the orbit path this consequence of Kepler's law means that the moment the engines are cut, the spacecraft will pass through that very location again in the coming orbital rotations. This is why capsules cannot just be launched from the ground into orbit with a catapult or cannon: The point where they leave the cannon barrel will again be part of the next orbit, which means that the orbital path intersects the Earth at some point.

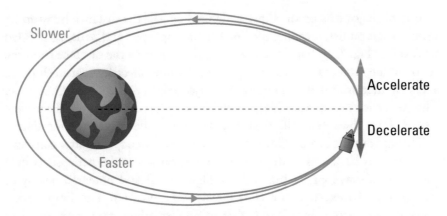

Fig. 3.10 An acceleration or braking manoeuvre at any point along an orbit mainly changes the shape of the orbit on the opposite end

When accelerating or decelerating in an orbit, by either applying thrust to nudge the spacecraft towards the direction of travel (*prograde*) or flipping the spacecraft around and letting the engines propel against the flight direction (*retrograde*), the same point likewise continues to be part of the resulting altered orbit (see Fig. 3.10). However, a highly unintuitive effect arises: Decelerating the vessel on its path means that it is still being pulled towards the central body, but now has less velocity to avoid it on the way down. Its new trajectory hence makes it fall around the body at a smaller distance while still missing it. The result is that the overall path and thus time for one orbital revolution gets shorter. Or to put it into other words: when slowing down in orbit, a spacecraft moving around the planet actually gets faster!

The other way around, by accelerating forwards, a spacecraft's orbit remains unchanged in the point of the thrust manoeuvre but it gains altitude on the opposite end—and reaches an orbit that makes it move slower around the central body.

This was more precisely formulated again by Johannes Kepler, in his second and third law:

A line drawn from the Sun to the planet covers identical areas of the orbital plane in identical periods of time.[1]
The squares of the orbit times are proportional to the cubes of the semi-major axes of the orbit ellipses.

---

[1]The second law is never really applied in this book, but it was historically important for the derivation of the third law.

The semi-major axis of an ellipse is defined as half the distance between its most distant points, meaning the line between apoapsis and periapsis divided by two (see Fig. 3.9). For a circular orbit, this is simply the distance from the centre of the planet to the orbiting spacecraft. It is noteworthy, that an elliptical and a circular orbit with the same semi-major axis have the same orbital time. The eccentricity of an orbit (describing how much it deviates from the perfect circular form towards an elliptical shape) does not affect its revolution time!

Throughout the orbital ellipse, the overall energy of a spacecraft is constant, as it converts back and forth between kinetic energy at the periapsis (where the spacecraft swoops by the body at high speed) and potential energy at the apoapsis (where the spacecraft moves rather slowly). The simple take-home message from these laws is that in smaller orbits, spacecraft complete a revolution around the planet more quickly and in larger orbits, they take a longer time to circle around. This rule of thumb will come in very handy when considering orbital manoeuvres, especially when trying to rendezvous with or avoid other objects in orbit.

Crunching the Numbers: Orbital Times of Spacecraft and Altitude of TV Satellites

To calculate the time an object takes to orbit once around its central body, Kepler's third law becomes convenient. In its basic form, it relates the orbital times of two objects, T_1 and T_2, with their orbital distances a_1 and a_2 (more precisely, with the semi-major axis of their orbital ellipse, which corresponds to the radius of a circular orbit):

$$\left(\frac{T_1}{T_2}\right)^2 = \left(\frac{a_1}{a_2}\right)^3 .$$

The purpose of this equation is to calculate either the radius or the orbital time of an object, when the other quantity is known. But this formulation is not, by itself, very practical, as it only allows these values to be calculated in relation to a second object orbiting around the same central body, like both Earth and Mars orbiting the Sun.

Thankfully, the equation can be reformulated once the orbit time and radius of a single orbiting object are known. It is rewritten in terms of the mass of the central body, no longer needing the values of two objects. Now it can be used to calculate an object's orbital time from its distance (and vice versa):

$$T^2 = \frac{4\pi^2}{G_N M} a^3 ,$$

(continued)

where G_N is Newton's gravitational constant ($6.67 \times 10^{-11} \, m^3/kg\,s^2$) and the central body's mass is M (In the case of Earth: $M_E = 5.97 \times 10^{24} \, kg$).

This formula makes it possible to calculate, for example, the time that the International Space Station takes to orbit once around the Earth, by inserting its altitude of about 416 km plus the Earth's radius of 6378 km as the value a. Solving for the orbital time T yields:

$$T_{ISS} = \sqrt{\frac{4\pi^2}{G_N M_E}(6794 \, km)^3} = 5575 \, s = 92.9 \, min.$$

The resulting orbital time of the ISS around the Earth is a little bit over 1.5 h.

TV satellites have a *geostationary* orbit, meaning that each of them permanently sits above the same spot on Earth. This makes it easy to align a satellite dish with such a satellite, as it stays in the same location in the sky all the time. But how can it be both stationary and at the same time fall around Earth fast enough to miss it? Since the Earth rotates around its axis once every 24 h, the satellite needs to be placed into a circular orbit that takes the same 24 h to complete. To calculate the orbital altitude necessary for it, the equation can be solved for a, with $T = 24 \, h = 86,400 \, s$:

$$a = \sqrt[3]{\frac{G_N M_E}{4\pi^2} T^2} = 42,227 \, km.$$

Hence, all TV satellites are flying in about 42,000 km distance from the centre of the Earth. Subtracting Earth's radius, this means an altitude of roughly 35,000 km above the surface.

It is not a hundred percent correct to assume, that an orbit is a perfect ellipse and that any object placed into an orbit will stay there forever. Vanishingly tiny amounts of remaining atmospheric drag and collisions with micrometeoroids gradually slow down objects, so that their orbits decay over time. Also, since the Earth is not a perfect sphere and other celestial bodies' gravity has a diminutive effect, the orbits elliptical shape is perturbed ever so slightly over time (especially circular orbits will not remain perfect, but turn into ellipses). Hence, a little push from an engine is required now and then to keep the spacecraft in the desired trajectory. The effect is more prominent in lower orbits, where the atmospheric pressure is still more substantial. The ISS needs to boost its velocity about once a month to maintain altitude, whereas geostationary satellites stay up for thousands of years without any corrections (although they would start to drift around the Earth, away from their intended location).

Orbital Hohmann Transfer

Transferring from one orbit into another is among the elemental flight manoeuvres that every spacefarer should master, as it is essential for reaching spacestations and other celestial destinations. The basic principle, illustrated in Fig. 3.11 with the example of two circular orbits, is quite easy: As mentioned before, accelerating forward at any point of an orbit raises the altitude of the opposite end of the flight path, whereas decelerating against the direction of motion causes it to lower. If the plan is to change from a low orbit into a higher one, the process starts by applying thrust for a sufficient duration until the opposite end of the orbit has reached the target altitude. Now following the elliptical transfer orbit, one has to simply wait for half a revolution for the spacecraft to reach the new highest orbital altitude. Upon reaching this point, another boost of forward thrust lifts the opposite end of the orbit to the desired altitude and turns its elliptical shape back into a circle.

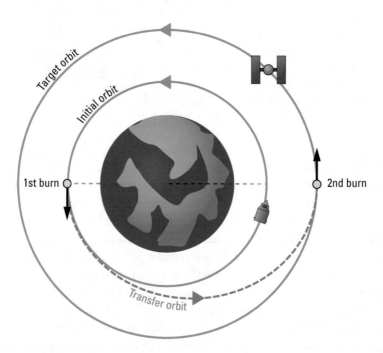

Fig. 3.11 Diagram of a Hohmann transfer. Starting from a low orbit, the spacecraft first performs a burn to change to a more elliptical transfer trajectory that touches the target orbit with the apoapsis. Once the target altitude is reached, a second burn circularises the orbit

This process is called a *Hohmann transfer*, named after the German rocket pioneer Walter Hohmann, who already proposed it in his 1925 book *Die Erreichbarkeit der Himmelskörper* (Translation: The Attainability of Celestial Bodies, Hohmann 1960), decades before the first rockets reached space.

The Hohmann transfer divides the orbit change into a couple of short bursts of engine activity that strongly alter the orbit only on one end. Hence, this transfer has a much higher energy efficiency to reach the target orbit than other methods with longer engine burn times. In comparison, using a small rocket engine that continuously provides thrust in forward direction and thus moving from the start to the destination on a path that is slowly winding outwards as a spiral uses more propellant to achieve the same result. Especially when the ratio between the original and target orbit radius is large, this difference can be quite significant: When going from low Earth orbit (400 km altitude) to geostationary orbit (35,000 km altitude), for example, a Hohmann trajectory is almost 80% more energy efficient than a spiral path!

If the target orbit is elliptical, this manoeuvre should not be started at any arbitrary point, but ideally begin by raising the altitude of the apoapsis, the point furthest from the Earth. To do so, the first acceleration step must be performed adjacent to the target orbit's periapsis, so that the transfer orbit's apoapsis matches that of the target orbit. If any other point along the orbit were used to start the transfer, it would not only be less energy efficient, but it would also no longer be truly a Hohmann method. The thrust would no longer be used to push the craft strictly in forward direction of travel, but in a heading oblique to the flight path.

The Hohmann transfer is uncomplicated, energy efficient and practical—but unfortunately also very slow: This type of transfer will always require at least half a revolution around the central body. While this timescale is still quite reasonable for manoeuvres involving orbits around the Earth, it becomes time consuming when planning manoeuvres from one planet's orbit to another. For example, a Hohmann transfer from Earth to Jupiter, where the Sun acts as the central body, would take half a Jovian year (which equals 11.8 Earth years). This results in 6 years of flight time! Human missions into the outer solar system will have to rely on more direct and fuel intensive flight plans, in which a considerably steeper transfer orbit is used and more fuel is expended both at departure and at the destination to slow down and match the orbital speed of the target.

Tilting the Orbital Plane

So far, orbital motion was treated as though it were confined entirely to a two-dimensional area, like a flat plane. Indeed, in the absence of additional forces acting on the spacecraft, beyond the gravitational pull of the central body, the physical principle of angular momentum conservation keeps an orbit perfectly aligned with the direction it was originally launched into. Since planets and other celestial bodies are not two-dimensional circles but three-dimensional objects, orbits can be established in all possible planes around them (like a ball of yarn).

The plane that the orbital ellipse lies in cuts through the centre of the planet at an angle called the orbital inclination. On Earth, this angle is measured against the equator, so that a craft travelling from west to east exactly above the equator (like geostationary satellites) is said to have a 0° inclination. Orbits over the poles have an inclination of 90°, and trajectories travelling from east to west (on so-called retrograde orbits) are denoted by inclinations above 90°.

To fully describe the orbit in three dimensions, an additional angle around the polar axis describes at which spot the orbital plane intersects the equator. It is referred to as the "Right ascension of the ascending node" (RAAN). Since Earth itself rotates, this angle is specified relative to a fixed point in the sky ("First Point of Aries", the point where the Sun is located in the sky at the spring equinox). Any change of either inclination or right ascension corresponds to a tilting of the orbital plane.

Unfortunately, a direct launch to the intended target inclination and right ascension is not always possible. A zero inclination orbit, for example, would require a launch pad directly on the equator (the closest one is the European Space Agency's Kourou launch site in French Guiana, 5° north of the equator). Likewise, if intending to fly from one spacestation to another, their planes might not match up.[2] If that is the case, it is necessary to perform a manoeuvre for tilting the orbital plane.

Figure 3.12 portrays this manoeuvre. Like before, this method is relying on the principle described by Kepler's first law, which states that the point where the plane change manoeuvre has concluded will again be part of the trajectory in future revolutions. To change from one orbital plane to another, the process starts by waiting until an intersection point is reached between the planes. At that moment, the spacecraft's rocket engines are used to accelerate

---

[2]The only ever flight from one spacestation to another, where Soyuz T-15 flew from Mir to Salyut 7 and back, did not require a plane change since both stations were intentionally placed into the same orbit.

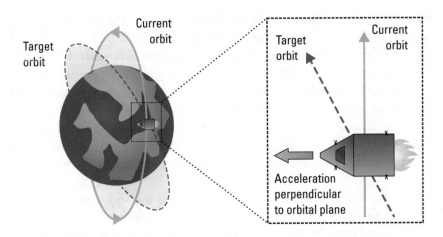

Fig. 3.12 Tilting the orbital plane: At the intersection point between current and target orbit, perpendicular thrust to the current plane tilts the trajectory (turning the spacecraft does not by itself change its direction of travel)

perpendicular to the original plane, so that the conceptual velocity vector starts tilting towards the new plane. With the correct timing, the spacecraft will continue its travel in the target plane with the intended inclination afterwards.

The implicit assumption in this manoeuvre is that the spacecraft's velocity can be changed instantaneously, meaning that the rocket engines can turn the direction of motion by multiple degrees in a split second. This assumption, unfortunately, does not hold true in reality: In the customary flight plan of a Soyuz spacecraft to the International Space Station, the plane change can take up to 2 min. In satellites and space probes, the small navigational thrusters likewise need several minutes for such a manoeuvre. Spacecraft are simply not able to "turn on a dime" at one point of the orbit, but have to perform an extended rocket burn close to the intersection point. As a modus operandi, the manoeuvre can be planned such that two thirds of the rocket burn time is spent before and one third after the expected intersection point. If this is insufficient to reach the target orbit in one go, the plane turn can be split up into multiple such manoeuvres over subsequent revolutions around the planet.

It should be kept in mind, that tilting of the orbital plane by more than a couple of degrees requires a massive amount of fuel: For the extreme case of a rotation of 90°, the entire kinetic energy of the orbital motion has to be gotten rid of, and at the same time re-acquired in perpendicular direction, to not plummet downwards. (Flipping an orbit around completely to move backwards in the same plane would be even more ridiculous.) For such a stunt, an equally large rocket would be needed that launched the spacecraft from the surface in the first place. Whenever possible, the launch site or initial orbit should be chosen to minimise inclination changes afterwards. If the launch site is placed such that launching into the target orbit is possible by going due eastward, an additional bonus in energy is awarded as the Earth's rotation gives the spacecraft a free boost. The Cosmodrome in Baikonur is located at 51.56° northern latitude, and when launching something straight eastwards from there, such as a supply transporter, it ends up in an orbit with 51.56° inclination. This explains why the International Space Station orbits in that plane.

A plane change using this manoeuvre can be too fuel expensive in some cases, so the gravity of other celestial bodies can be used to help with the inclination change (see Sect. 3.5). Once the orbit is coplanar with the target, worrying about the three-dimensional nature of orbital motion is no longer required, and all thinking can return to two-dimensional terms.

Rendezvous and Docking

The previously discussed manoeuvres—Hohmann transfer and orbital plane change—allow a spacecraft in principle to reach any orbit of choice (provided it carries a sufficient amount of propellant). In order to fly to a spacestation and dock there, however, it is not enough yet to be in an orbit with the same altitude and inclination, as also the angle of rotation around the orbit, called the *phase angle*, needs to match up. For example, if a spacecraft were to launch into an orbit with the altitude and inclination of the International Space Station without any regards to its position on that orbit, the orbit times of the ship and the station would be identical according to Kepler's third law. They would simply circle around Earth indefinitely at a fixed distance from one another and would never rendezvous.

It is likewise not a good idea to point a spacecraft right towards the intended target vessel, boost ahead, and expect to encounter it on a straight line: Since orbital motion occurs in ellipses, this would basically guarantee never to arrive at the intended destination. In science-fiction depictions of spaceflight, this is one of the most common physics goofs!

Instead, the task is now to combine the previously discussed physical principles and manoeuvres in a smart way not only to arrive at the correct orbit, but also with the right timing. The typical, well established strategy on how to achieve this was presented by the astronaut and engineer Buzz Aldrin in his PhD thesis "Line-Of-Sight Guidance Techniques For Manned Orbital

Rendezvous" published in 1963 (he continued to become famous for walking on the Moon as the second person after Neil Armstrong).

Kepler's third law forms the starting point of this procedure, as it relates the semi-major axis of the orbit with the orbital period (see Fig. 3.9): When flying on a lower orbit than the target object, the semi-major axis is shorter and consequently the vessel circles the planet faster. As a result, the lower vessel passes under its target after a while. Vice versa, by travelling on a higher altitude orbit than the target, the orbital period is longer and the pursuing vessel falls behind. The procedure for launch towards a spacestation from the ground begins by waiting until the station flies directly overhead. At that time, the spacecraft is launched into a somewhat lower orbit (which, at the same time, should be coplanar with the orbit of the target, to avoid expensive plane-change manoeuvres). The station, in the mean time, gains a head start.

A spaceship, having reached the lower orbit, first undergoes basic house-keeping tasks such as system checkouts (like verification of the correct function of thrusters and the life support system). Once these are complete, the first half of a Hohmann transfer towards the target orbit is carried out. In this phase, the apoapsis, i.e. the point furthest from the Earth is lifted to coincide with the altitude of the target object. The second half of the Hohmann transfer, however, is delayed, so that the periapsis remains on the level of the initial, lower path. As a result, the semi-major axis of the orbital ellipse in this transfer orbit remains smaller than that of the target object and with every revolution around the Earth, the distance to the target decreases.

Eventually, the approach will lead to a point in which the destination would be caught up with or even overshot in the next revolution; now it is time for the next engine ignition. The engine burn starts the second phase and just like the second half of a Hohmann transfer, it is performed at the apoapsis of the transfer orbit and raises the periapsis. The orbit is not yet completely lifted to match up with the destination, but the distance between the periapsides is halved, thus reducing the difference in orbital times and slowing down the approach.

Figure 3.13 visualises the complete sequence and shows the orbital motion of the approaching spacecraft from the point of view of the target object, meaning that all objects are rotated around the Earth's centre at the orbital speed of the target. As a result, the target remains in one fixed spot, while an elliptical orbit going up and down between peri- and apoapsis altitudes shows up as a spiral.

For a long time, flights to the International Space Station employed precisely this method, both by Soyuz and Space Shuttle. Since safety distances were considered of utmost importance, it took about 20 orbital revolutions, or one and a half days, from launch to arrival at the station. Gained experience

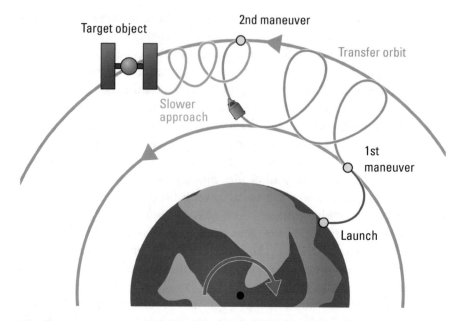

Fig. 3.13 Illustration of an orbital rendezvous manoeuvre, shown from the viewpoint of the target object, so that all other objects are rotating in relation to it (the rotational axis is marked by the black dot on the globe): The approaching spacecraft is slowly closing in on an elliptical orbit that cycles back and forth between the two altitudes shown in grey. Shortly before reaching the target, its periapsis is raised, and thus the difference in orbital time is reduced to slow down the approach. Eventually, a direct final approach is taken

and improved navigation technologies have nowadays increased manoeuvring precision, which permits flights to the ISS with much smaller safety margins. In 2012, the flight duration was reduced to four orbits, or to about 6 h, by choosing a launch trajectory that brings the spacecraft into a higher initial orbit, much closer to the doorstep of the ISS. This shorter flight duration is advantageous for many reasons, such as for the transport of refrigerated or perishable supplies. It is also a welcome change from the spacefarer's perspective because the Soyuz capsule does not contain a very convenient toilet. A further reduction to only two orbits has been tested with uncrewed Progress transporters in 2019, and will probably become the norm for all missions to the ISS.

The boost to lift the periapsis can be repeated multiple times, until the spacecraft has arrived sufficiently close (usually around a 100 m) to the target. In the simplest version of this manoeuvre, the spacecraft closes in towards the target from below. Since it is in the upwards motion of the last transfer orbit, the encounter will normally happen so that the station is above the incoming vehicle. Alternatively, it is possible to fly around the spacestation by entering an orbit that is just slightly more elliptical (but has the same semi-major axis as the target). The arriving spacecraft then slowly circles around the target (like two dogs chasing each other's tail), as the two vessels' velocities vary throughout one orbital revolution. This can be useful, for example, to visually inspect the destination, or to align with a suitable docking port on a spacestation.

Once the incoming craft is aligned for docking, any remaining velocity difference is eliminated. Both spacecraft are now in nearly identical orbits, and the last leg of the flight (the *final approach*) towards the chosen docking port is performed in a linear manner, finally neglecting the curvilinear properties of orbital motion. Any deviation away from the straight approach line can be compensated with the control thrusters.

The actual docking to the station can be achieved with a number of different mechanisms. Currently, the following four types of docking ports are available on the International Space Station:

- The adapter called *Common Berthing Mechanism* (CBM) is what most uncrewed commercial transport craft applied so far. The strategy to dock with the ISS is simply to fly close enough to the station (about 5 m distance) and hold position until the station's robotic arm grabs the craft and places it on a free docking port. This adapter is also found between the modules of the ISS. The station itself was assembled with the help of the robotic arm mounted in the Space Shuttle's cargo bay. The disadvantage of this method is that it demands active participation from the docking target and its robotic arm (Fig. 3.14a).
- The *Androgynous Peripheral Attach System* (APAS) is a docking standard originally developed in cooperation between the United States and the Soviet Union for the Apollo-Soyuz Test Project. It continued its legacy in the Space Shuttle docking ports and connects the Russian and the American part of the International Space Station. Also the docking port system on the Chinese Shenzhou spacecraft is designed to be compatible with APAS. The *soft dock* procedure it employs consists of two stages: First, the spacecraft approaches with low speed (few cm/s) and brings a docking ring fitted with shock absorbers into contact with the docking partner. Next, when oscillations have dampened down, the ring hydraulically pulls both craft in, and only then a solid, airtight connection is established (Fig. 3.14b).
- The International Docking Adapter (IDA) standard is a further evolution of the same design and employs the same docking process as APAS. In addition to the transfer of crew and electricity, it also foresees the ability to pump, e.g.

a)   b)

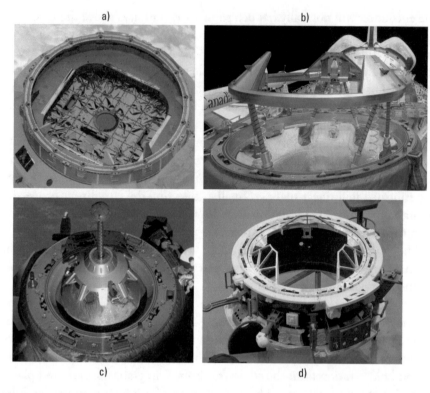

c)   d)

Fig. 3.14 The four varieties of docking ports available on the International Space Station: **(a)** The CBM standard connects station modules to each other and enables spacecraft to be attached by the station's robotic arm. **(b)** The APAS system of the Space Shuttle orbiters, with the active docking ring extended. **(c)** The Russian SSVP system of Soyuz and Progress spacecraft. **(d)** The new International Docking Adapter standard that is intended for commercial spacecraft (Images: NASA & B. Ganse)

fuel and oxygen from one docking partner to the other. On the space station, there are already two such ports to serve current and future commercial space vehicles. Dragon 2, Orion and Starliner spacecraft are equipped with this docking port (Fig. 3.14d).

- Quite an opposite approach is taken in the *hard dock* concept underlying the Russian SSVP (*Sistema Stykovki i Vnutrennego Perekhoda*, mechanism for docking and internal transfer) port, which was also present on the European ATV transporter: Here, a typical closing speed of around 10 cm/s is required to immediately create a solid connection between the docking partners, as the approaching spacecraft latches into the docking port with the help of its momentum. After successful docking and leak checking, the probe extending from the centre of the docking port is retracted out of the way so that persons can pass through it (Fig. 3.14c).

All four docking ports have in common that at the end of their docking process, a mechanically stable and airtight connection forms between both vessels. Multiple layers of rubber o-rings or sealing gaskets are held tight by robust holding clamps to ensure that no gas can leak out into space.

3.4   Flying to the Moon

The Moon is, first and foremost, an ordinary satellite of Earth, that circles in a high orbit in the course of 1 month. Consequently, flying to the Moon does not fundamentally differ from the ideas before: The travel starts by launching into a low Earth orbit, followed by a Hohmann transfer so that a rendezvous with the Moon results.

The Moon's path around the Earth takes place in a plane that goes through the centre of the Earth and is tilted 28° away from the equator, so the journey should ideally start with a launch into an orbit with that inclination. Optimal launch sites are themselves located at a latitude of either 28° north or south. At these latitudes, the launch direction is due eastwards, so that a maximum amount of momentum is gained from Earth's rotation. And lo and behold: NASA's launch complex is located in Florida and its latitude is exactly 28° north. Once a month, the Moon passes directly through the zenith there.

However, the launch day does not need to be on the date the Moon transits exactly overhead. To avoid a costly plane-change manoeuvre, it is sufficient for the launch site to lie inside the orbital plane. In general, the rotation of the Earth makes a launch window with the correct inclination and right ascension available once a day at 28° latitude, and twice for launch sites closer to the equator.

When starting to get close enough to the Moon, its gravity begins to dominate over Earth and the trajectory is suddenly no longer determined by Earth as a central body! In fact, it is now possible (at least to some amount) to ignore the gravitational pull of Earth and switch the navigational point of view to consider the Moon as the central reference mass. From the perspective of the Moon, it looks like a spacecraft is approaching on a free fall trajectory from far, far away. Since the vessel comes from outside of the gravitational field of the Moon, it carries enough energy to escape the pull on the other side. The craft thus has a different type of flight path than the previously considered orbits: The track does not have the shape of a closed ellipse, but forms a hyperbola that looks like a boomerang (green line in Fig. 3.15). This path, like an ellipse, swings around the planet but does not curve back to meet itself. Instead, as

Braking manoeuvre

Fig. 3.15 When approaching the Moon from a faraway distance, a vessel follows a hyperbolic trajectory, unless a braking manoeuvre is performed to enter an orbit (red). The transition from a hyperbolic towards an elliptic trajectory is indicated by the yellowish trajectories

it recedes, it changes more and more into a straight line. In other words, the apolune point (furthest orbital distance from the Moon) is located at infinity.

On this trajectory, the spacecraft first gets closer to the Moon, encounters its closest approach point (the perilune) and then symmetrically moves off into empty space again—unless the route is altered in a navigational action or it is coming in too low and intersects the lunar surface at the closest point.[3]

To enter a stable orbit around the Moon thus demands a manoeuvre. But which one? Fundamentally, this is a repetition of the same process as before: By decelerating at one point of the orbit, the orbital altitude at the opposite end of the path is lowered. In this specific case, the goal is to decrease the orbital velocity sufficiently, so that the spacecraft is no longer flung out of the Moon's gravitational pull. The resulting form of the flight path will again be an ellipse, so it is best to perform this manoeuvre at the periapsis. The required burn time can either be calculated using Kepler's third law (in which a hyperbola will get a negative value for the semi-major axis), or by using a simple rule of thumb. When approaching the periapsis point, the spacecraft's altimeter shows the altitude over the Moon's surface to be first decreasing, then levelling out, and then rising again, if the speed is too high to be bound by the Moon's gravity. The rocket engine needs to run until the speed has been reduced to fit a circular orbit and the altitude stops changing. If the spacecraft continues slowing down beyond that, the opposite end of the orbit will drop down and might end up touching the Moon's surface. To be on the safe side, it is advisable to always confirm the burn time with calculation.

---

[3]This could also be called a braking manoeuvre of some sort, even though it would be a rather uncomfortable way of slowing down.

Starting from a stable lunar orbit, preparations for landing can begin by shedding all remaining orbital velocity. Without an atmosphere to utilise parachutes in, the spacecraft's rocket engines reduce the speed before touchdown to a few metres per second. Luckily, the Moon's gravitational pull is notably weaker than Earth's, meaning landings were already possible with the rocket technology of the 1960s.

To return from the Moon, the same procedure is executed backwards: The spacecraft launches into lunar orbit and extends the trajectory until the furthest point is located in infinity. Ideally, the gravity well of the Moon is left in the opposite direction to the Moons orbital motion around the Earth, so that the spacecraft falls right back into the Earth's atmosphere (similar to the path shown in Fig. 3.17).

3.5   Travelling Towards Other Planets

Travel from one planet to another inside the solar system is very similar to the rendezvous manoeuvres discussed before. Again, the goal is to adjust the flight path in such a way that the target object is encountered and to perform a suitable deceleration manoeuvre to enter an orbit around it.

The most important difference is that the central object, in respect to which all flight paths are being considered, is no longer the Earth, but the Sun—and thus all orbital radii are substantially more extensive and flight times therefore considerably longer. For example, the time for Earth to orbit once around the Sun is known as a *sidereal* year: 365 days, 6 h, 9 min and 10 s (after this amount of time, the Earth is again in the same spot on its orbital path).

When a Hohmann transfer is followed to get from planet to planet, a travel time of half a transfer ellipse needs to be accounted for. Since this orbit lies between the start and destination bodies' paths around the Sun, flying the transfer trajectory takes between half a year of the departure planet and half a year of the destination planet (see the example calculation of the flight time to Mars below). This can be an enormous time for the outer planets of the solar system. While the flight duration is less of a concern for space probes (as they require neither food, water nor entertainment), bringing humans to planets and moons beyond Mars or to targets in the asteroid belt would demand facilities for spending months and years en route.

Like in the rendezvous procedures around Earth, the departure date for the Hohmann transfer should be chosen so that the transfer ellipse actually ends up at the target planet. If the departure and target planet are not in a suitable alignment for it, the journey would just end up touching the orbit of the target, but not necessarily very close to the planet. Some additional months of waiting time might be needed until the alignment is right.

Crunching the Numbers: Travel Duration to Mars

Following the prescription for a Hohmann transfer, the trajectory for a flight from the Earth to Mars is half an ellipse, whose major axis corresponds to the sum of the orbital radii of Earth and Mars.

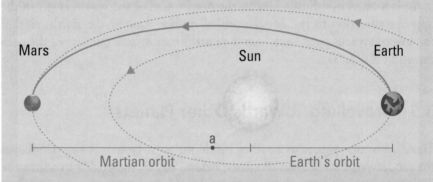

The semi-major axis, a, divides the long axis of the transfer ellipse into two equal halves:

$$a = \frac{R_{\text{Earth}} + R_{\text{Mars}}}{2} = \frac{149\,\text{million km} + 227\,\text{million km}}{2} = 188\,\text{million km}$$

The earlier calculation of satellite orbits and the orbit of the International Space Station around the Earth got the period of an object on an elliptical flight path from a reformulation of Kepler's third law. Exchanging the mass of Earth with that of the Sun ($M_\odot = 2 \times 10^{30}$ kg), the same formula can be reused to calculate the fight time to Mars:

$$T = \sqrt{\frac{4\pi^2}{G_N M_\odot} a^3}.$$

(G_N is again Newton's gravitational constant, with a magnitude of $6.67 \times 10^{-11}\,\text{m}^3/\text{kg}\,\text{s}^2$.)

Inserting the value of a from above, the time for one full transfer ellipse flight results to:

$$T = 44{,}344{,}356\,\text{s} = 513\,\text{days}.$$

A one-way flight to Mars however does not go through the full ellipse, but arrives at its target after half of this time. Hence the resulting flight time from

(continued)

planet to planet is half of the above value, 256 days. For a return to Earth, an equivalent trajectory is followed.

In this somewhat coarse derivation, the planets' orbits were assumed to be circular, their differing orbital planes were neglected and the mass of the Sun was approximated to a round number. Yet, if this estimate is compared with reality, it turns out that the calculation is only about 1% off, as the Mars rover Curiosity took 253 days to arrive at Mars.

A Hohmann transfer is not the only way to perform an interplanetary flight. As described in Sect. 3.3, it just happens to be the cheapest, since it prescribes the lowest amount of velocity change to reach the target orbit. If a budget for a more massive rocket containing more fuel is available, flight durations can be shortened considerably, thus saving on food, water, oxygen and other supplies for the journey. Physics (as described in the rocket equation, see Sect. 2.3), however, has the habit of making things difficult: Every additional kilogram of propellant or payload mass taken to orbit demands more propellant to do the lifting, which again needs more lifting capability...In the end, a journey that is 10% faster might require a rocket that is three times bigger! Travelling to Mars is an economically unfavourable endeavour to begin with and does not get any cheaper with exponentially growing fuel consumption that a reduction of travel time would entail.

Most plans for flights to Mars are based on the Hohmann trajectory for budget reasons, resulting in a mission duration of about 550 days for the entire round trip (including a month-long stay on Mars).

Upon arrival at the target planet, entering into orbit works as previously described for the Moon, by using rockets to slow down at the periapsis until a closed elliptical orbit is attained. If the target body has an atmosphere (all planets in the solar system, with the exception of Mercury do, whereas dwarf planets and moons typically do not), it can be used to save fuel for the deceleration. Instead of performing the orbit insertion with rocket engines, air friction is used for braking by slipping low enough into the atmosphere. This is called an *aerobraking* or *aerocapture* manoeuvre. It is immensely important to know about the atmospheric density profile of the planet beforehand: Entering too deep into dense strata of the atmosphere, heating can become too intense, much like in an atmospheric reentry (see Sect. 2.2). On the other hand, if only elevated, tenuous parts of the atmosphere are reached, this can result in insufficient deceleration, and the spacecraft continues to travel on a hyperbolic trajectory, failing to enter a closed orbit. Once aerobraking has lowered the apoapsis to an adequately low altitude, a short acceleration burn raises the periapsis out of the atmosphere (except if the intention is to keep lowering the orbit and land on the planet).

The Swing-by Manoeuvre

If the solar system were to consist only of Sun, Earth and one other planet, the treatise of possible flight paths would be complete here. In reality, however, the Sun's neighbourhood contains eight planets, multiple dwarf planets, countless moons, asteroids and comets. Also, the previous assumption of a central body at rest in some fixed frame of reference that keeps all other objects in its grip, is never quite true either. Moons orbit around planets, that revolve around the Sun and the entire solar system circles around the center of the galaxy while the whole universe is expanding! Could there be flight trajectories which have been ignored in the prior deliberations that were limited because only two or three bodies were considered at a time? How can these be of benefit?

Multiple sorts of many-body trajectories exist, and some are beating the Hohmann transfer by a wide margin in terms of fuel efficiency and travel speed! Unfortunately, many of them involve complex gravitational effects of multiple sources and are neither easy to illustrate, nor to calculate (NASA itself has tasked a supercomputer the size of a basketball court to find such mission flight plans, as pen and paper are no longer helpful here). Some types of trajectories are still easy enough to figure out and put into use.

In a *swing-by*, also known as the gravitational slingshot manoeuvre or flyby, the gravitational pull of a body is exploited while flying past it without entering into a stable orbit. Just like described in Sect. 3.4, the celestial body (be it a star, planet, asteroid or moon) is approached from a quasi-infinite distance, following a hyperbolic flight path (see Fig. 3.16). Due to the pull of gravity, the flight path curves towards the planet. The highest velocity is reached at the point closest to it (the periapsis) and the craft then hurls back towards infinity on the other side, if no thrust is applied.

Previously, when entering into lunar orbit, it was exactly at the periapsis where rocket engines were ignited to slow down the orbital motion into an elliptical shape. But this time, in a swing-by, the spacecraft is never expected to enter a stable orbit, instead making use of the slingshot effect of the planet.

While this whole procedure appears completely symmetric from the point of view of the swing-by planet—the spacecraft comes in from one side, quickly swooshes around and buzzes away into the other direction—a marvellous trick is performed when looking at the solar system as a whole: Without using a single drip of rocket fuel, the spacecraft's direction of motion has been

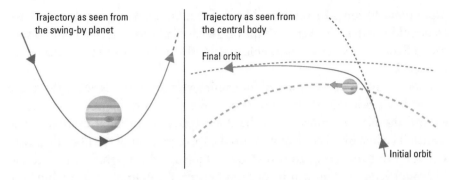

Fig. 3.16 Illustration of a swing-by manoeuvre: From the point of view of the swing-by planet, the spacecraft simply flies a hyperbolic flight path, without any use of propellant. From the perspective of the central star, however, the spacecraft gains momentum in this flyby

noticeably altered (nearly 180° turns are possible!). As an additional bonus, the spacecraft can get a sizeable boost in velocity, as the swing-by planet itself has an orbital speed around the Sun. This effect is largest if the craft gets flung out along the direction of orbital motion of the planet, so that their velocities add up most.

Conservation of momentum, the principle underlying all propulsion in space (see Sect. 2.3) naturally continues to be valid, also in this stunt. In other words, a velocity gain of the spacecraft during the swing-by must be counterbalanced by a miniscule loss in velocity of the planet. But due to the enormous masses of planets, there is no reason to be concerned. It would require billions of spacecraft swing-bys to alter the orbital period of Jupiter by even a second.

The heavier the planet that is used for this manoeuvre and the closer the spacecraft swoops around it, the stronger the deviation of the flight path will be and the more energy can be gained in the swing-by manoeuvre. The gas giant Jupiter thus acts as something like the central traffic hub of the solar system, both for missions to the inner and the outer planets … and beyond.

The two Voyager probes are excellent examples of the power of swing-by trajectories. They launched on a *Grand Tour* of the solar system in 1977, where a fortuitous alignment of Jupiter and Saturn allowed the probes to combine gravity slingshots from these two gas giants. Voyager 1 was the first spacecraft ever to leave the solar system and is currently the furthest human-made object, travelling in interstellar space towards the stars. Voyager 2 performed two more swing-bys on Uranus and Neptune and is likewise headed out of the solar system. In total, through multiple swing-bys, each of the Voyager probes got more than 20 km/s in velocity. This is far more than they could ever have achieved from launch, even with the biggest available rocket! (For comparison: even a Saturn V rocket was only able to gain 10 km/s in velocity; see calculation in Sect. 2.3.)

A swing-by is not solely useful for gaining velocity, but it is equally possible to use it for slowing down: To this end, the process of approach and curve around the planet simply has to be done against the orbital motion of the planet. The hyperbolic trajectory leads the spacecraft out of the planet's gravity well following the retrograde direction, ending up with a significant reduction in orbital velocity. This can be of great benefit, for example, for reaching the inner planets of the solar system. To transfer into an orbit close to them, a significant deceleration from Earth's orbital velocity around the Sun is essential. For a flight to Venus, it can be more fuel efficient to first fly out to Mars and undergo a decelerating swing-by to lower the perihelion, than to reduce speed with rocket power alone and travel to Venus directly. (Even though the orbits of Venus and Mars are approximately the same distance from Earth's, Venus sits much deeper inside the Sun's gravity well. To descent to its level, plenty of speed needs to be shed first, as the centrifugal force otherwise keeps the spacecraft on a wider orbit.)

On its flight to Mercury, the MESSENGER probe combined multiple swing-by manoeuvres. It first travelled from Earth to Venus to decelerate with two consecutive swing-bys and then encountered Mercury three times without entering an orbit. Only after all these flybys it arrived at Mercury with a low enough velocity to reach a stable orbit.

Anecdote

The Japanese Akatsuki space probe was supposed to enter orbit around Venus in December 2010 after a 6 month flight. Due to an engine problem, the on-board computer powered down the propulsion system as an emergency measure and the spacecraft started tumbling. Without having sufficiently slowed down, Akatsuki failed to enter orbit and performed an unplanned swing-by around Venus. It suddenly found itself in a solar orbit that led it very close (0.6 Earth orbital radii) to the Sun, exposing it to higher heat loads.

Only 5 years later, in December 2015, its orbit fortunately again came close enough to Venus to successfully enter into an elliptical track around it. The accidental swing-by might just as well have led to a trajectory, from which Venus would have been out of reach. There were even fears that the probe would melt in close proximity to the Sun!

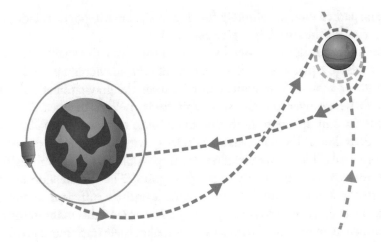

Fig. 3.17 The flight path of the Apollo missions was a circumlunar free return trajectory. In case the engine ignition to enter lunar orbit had been missed, the flight path would have resulted in a direct reentry back into Earth's atmosphere

Another special sort of retrograde swing-by manoeuvre is the *free return* trajectory, which was implemented in the Apollo missions (Fig. 3.17) and is considered in plans for future missions to Mars. Like before, the target body is approached on a flight path against the orbital motion, so that a decelerating swing-by ensues. If everything goes well, a braking manoeuvre is then carried out to enter orbit around the target body and to eventually land on it. In case any problems occur during the flight, including such severe events as the complete loss of control of the spacecraft (e.g. due to engine malfunction or a fuel leak), the orbital insertion can simply be skipped. The trajectory is set up thereupon to result in a retrograde swing-by, in such a way that the spacecraft loses most of its orbital momentum. It plummets back towards the central body and reenters its atmosphere without any need for additional manoeuvres.

It is a very welcome safety feature for crewed missions and it was of enormous advantage on the Apollo 13 flight, where the explosion of an oxygen tank rendered the command module unusable.

Lagrangian Points

Both the Sun and the Earth have a gravitational pull on every object travelling around them. This raises the question: Is there a point between the two bodies, in which their gravitational forces would exactly cancel out, so that a spacecraft could be parked there indefinitely, floating freely? Such a spot does indeed exist and it is called the first *Lagrangian point*, L1.

This point is, however, not in equilibrium: Even the smallest deviation earthwards (by pressure of the solar wind, the gravitational attraction of Neptune or a hit of a micrometeorite) causes the gravitational pull of the Earth to be stronger and the spacecraft finds itself to be on a direct path towards the atmosphere from that moment on. A comparably small deviation in the direction of the Sun causes the solar gravity to dominate, drawing the craft sunwards. These tiny variations are impossible to avoid, so thrusters are mandatory to stay put at this Lagrangian point. The thrusters do not need to be strong at all, a puny little ion engine is entirely sufficient to remain at L1. In practice, spacecraft typically slowly circle around this point in the plane perpendicular to the Sun–Earth axis, as it makes their track more predictable and allows multiple craft to share the same location.

The L1 point of the Sun–Earth system is popular with satellites observing the Sun, such as the Solar and Heliospheric Observatory (SOHO). The solar wind measuring Advanced Composition Explorer (ACE) and the gravitational wave observatory Laser Interferometer Space Antenna (LISA) also found their place there. In principle, a similar Lagrangian point exists between every pair of orbiting celestial bodies. In the case of most moons, however, the gravitational disturbance by the Sun is so strong that the point is essentially unstable.

As implied by the name, L1 is not the only point, in which a spacecraft can be parked force-free, but there are 4 more: The Lagrangian points 2 and 3 (known as L2 and L3) are also located on the axis formed by the two participating celestial bodies. One is situated beyond the orbit of the smaller body and the other one is on the opposite side of the system (Fig. 3.18). In these two points, the gravitational forces of both bodies do not cancel out, but pull in the same direction. They add up, as if they were one combined, heavier object, so that circular orbits of L2 and L3 have the same orbital periods as the smaller body has around the central one. The result is that spacecraft placed in these points rotate permanently collinear with the two bodies. These locations do not always exist: If a tiny moon orbits a massive planet, the points are located inside the moon and are thus impractical. Just like L1, these points are not keeping objects themselves, but call for stabilisation with thrusters.

Yet, the L2 point of the Sun–Earth system is of high practical relevance: Since it is located directly in the shadow of Earth, it is the perfect location to place telescopes, as disturbing light from the Sun and scattering from the Earth are always arriving from the same direction and thus easily shielded away. The James Webb Space Telescope, the next generation follow-up to the Hubble Space Telescope, is aiming for precisely that spot. Meanwhile, L2 of the Earth–

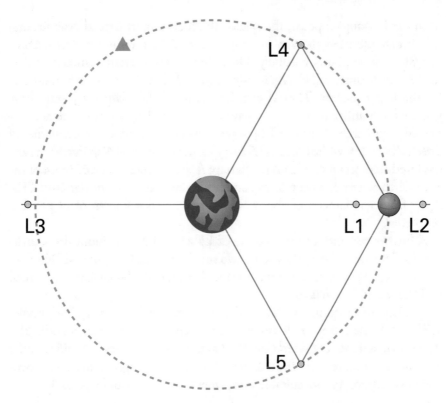

Fig. 3.18 Locations of the five Lagrangian points in a system of two orbiting celestial bodies. L1, L2 and L3 are located on the axis of the two bodies, while L4 and L5 are forming equilateral triangles with both bodies

Moon system is circled by the Chinese Queqiao satellite, which provides radio links to the back side of the Moon and works as a radio telescope. L3 of the Sun–Earth system, on the other hand, has never been the destination of spacecraft so far. Since it is located behind the Sun, communication with Earth would be quite difficult (it would, however, be the perfect place for a mad scientist to construct a secret space lair).

Finally, the L4 and L5 points each form an equilateral triangle with both celestial bodies in the same orbital plane, one located ahead of the smaller body and the other one trailing behind. It is not as directly obvious why these points would be stable, since the gravity of both bodies is pulling in different directions. Since the entire system is rotating, a dynamical balance of the forces is established here: If the spacecraft moves off the precise location of L4 or L5, its change in orbital velocity (and angular momentum) around one body is cancelled by the gravitational force of the other. Instead of being pulled away from the Lagrangian point, the spacecraft enters a bean-shaped orbit around it, without the need to even use any thrusters! As a result, objects become caught in these points naturally. Dust clouds and asteroids can accumulate in them over time, called *Trojan* asteroids, and they do so more numerously for the larger planets. The L4 and L5 points of the gas giant Jupiter have gathered a number of small companions, and in this case they are actually named after heroes from the Trojan wars: Asteroids at L4 carry the names of Greek characters, while those at L5 carry names from Troy. Also Earth's gravity has lured an object into a Lagrangian trap: A single Trojan asteroid in L4 of the Sun–Earth system has been found and been assigned the identifier 2010 TK$_7$. Apart from its diameter of about 300 m, not much is known of its physical properties.

A further curiosity of the stability of L4 and L5 can be found in Saturn's moons: The moon Tethys holds the two smaller moons Calypso and Telesto in its Lagrangian points as they orbit around Saturn, while Dione is accompanied by Helene and Polydeuces.

The Lagrangian points are based on the gravitational influence of two bodies at a time. If the forces of all moons, planets and the Sun (or even additional stars) are taken into consideration, the Lagrangian point location and stability becomes even more complicated. Yet, to a reasonable approximation, a two-body assumption gives a suitable idea, where spacecraft can be parked.

The MEGA Manoeuvre

Section 3.3 described that tilting of the orbital plane is very fuel intensive, but certain missions might include unavoidable changes of the orbital plane. Most flights from one spacestation to another would, for example, demand such a manoeuvre, unless they have specifically been placed in the same plane. So what can be done to make a plane change less fuel expensive? This is where the gravitational influence of the Moon comes into play, with the *Moon–Earth Gravity Assist* (MEGA).

To start with, this requires at least one end of the orbit to be raised, which is still a relatively cheap manoeuvre. The important part is to get sufficiently close to the lunar orbit for the craft to feel the gravitational influence of the Moon—and use its pull to alter the inclination angle. Instead of approaching the Moon within its orbital plane around the Earth, the trajectory is chosen such that the point closest to the Moon is clearly above or below that plane. The force of the Moon's gravity then acts on the spacecraft, turning the direction of travel.

The gravitational influence of the Moon does not have to be particularly mighty for this manoeuvre to work. If time is not a limiting factor (for example, in an uncrewed vessel) the distance to the Moon can remain quite wide, and only a couple of degrees of tilt is achieved on each approach. The method is available once per month, when the Moon is in a suitable location. Multiple months of flight time would probably be too long for missions carrying humans, so crewed flights would involve a slightly more fuel expensive path that leads much closer to the Moon's surface to reach the desired target plane in a single pass.

Such a manoeuvre is not limited to Earth and Moon but can likewise be accomplished with the help of other celestial bodies. The Cassini probe explored Saturn's moon system and performed a number of gravity assists around the moon Titan, changing its orbit from the ring plane of Saturn towards an orbit over the poles, by a whopping 75°.

Overall Mission Planning

The manoeuvres described in this chapter never occur in isolation, but are combined and sequenced for the complete flight plan of a space mission. In many cases, additional effects such as gravitational influence from other celestial bodies, pressure from the solar wind or remaining atmospheric drag need to be taken into account, meaning that the simple geometrical assumptions that underlie most of the deliberations only work as rough approximations. To actually have a reliable and predictable idea of the flight path, computer simulations (or a large squad of mathematicians with slide rules) are required to experiment with alternative routes, optimise fuel use, find ideal flight times and prepare for emergency contingencies.

NASA's software for this purpose, the *General Mission Analysis Tool* (GMAT) is an open source project that can freely be downloaded from the website linked in Fig. 3.19.

From a navigational point of view, all essential knowledge about spaceflight is now concluded, but there is still far more to keep in mind, as the next chapters will describe.

Fig. 3.19 NASA's General Mission Analysis Tool is available for download (http://gmatcentral.org/) and can be used to plan manoeuvres and trajectories for real and fictitious missions in the solar system

References

Aldrin, E. E. (1963). *Line-of-sight guidance techniques for manned orbital rendezvous.* Massachusetts Institute of Technology, Cambridge, MA. https://doi.org/1721.1/12652

Hohmann, W. (1925). *Die Erreichbarkeit der Himmelskörper.* München: Verlag Oldenbourg. ISBN 978-3-486-23106-5.

Hohmann, W. (1960). *The attainability of heavenly bodies.* NASA Technical Translation F-44, Washington, DC.

4

Daily Life in Space

Contents

The first chapters of this book covered engineering and physics. Now, to go one step further, this chapter discusses daily life on board a spaceship or a spacestation. There are many differences to living on Earth which may come as a surprise!

4.1   Orientation Without Up and Down

The main difference in every day life compared to living on Earth's surface is weightlessness. Humans are used to seeing situations and environments in only one perspective, with a well-defined orientation of gravity. Our brains primarily know it that way. In spaceflight, it takes several weeks to fully adapt to the new properties of orientation. Learning how to position oneself in

© Springer-Verlag GmbH Germany, part of Springer Nature 2020
B. Ganse, U. Ganse, *The Spacefarer's Handbook*, Springer Praxis Books,
https://doi.org/10.1007/978-3-662-61702-1_4

space next to the ceiling, that is not really a ceiling in this context, but a wall and the floor at the same time, is challenging. Photos and films of crews entering spacestations for the first time usually show all of them in the same orientation, which means their heads all point in the same direction. When they trained for their mission in a mockup of the station on the ground, gravity always gave them a clear indication of what is up and what is down. Later in the missions, videos give a totally different picture: the volume of space is fully used by people lingering in all corners, and with their heads pointing in arbitrary directions. Many people struggle finding their way around the ISS when on board for the first time, as they do not recognise where they are, despite knowing the station by heart from their extensive training on the ground (Fig. 4.1). Orientation is most difficult at intersections of modules. For that reason, labels and arrows are attached almost everywhere to help (Fig. 4.2). Painting the station modules in distinct colours would likewise help with orientation (in the Salyut 7 spacestation, for example, the walls were white, beige or apple green, depending on their orientation). Video tours of the ISS, see, for example, the one by astronaut Sunita Williams (QR-code in Fig. 4.3), may illustrate this problem. These tours are also a great way to learn about the station and its configuration.

In weightlessness, pushing oneself off from a wall results in linear floating into one direction. After a while, it starts to feel normal that locomotion

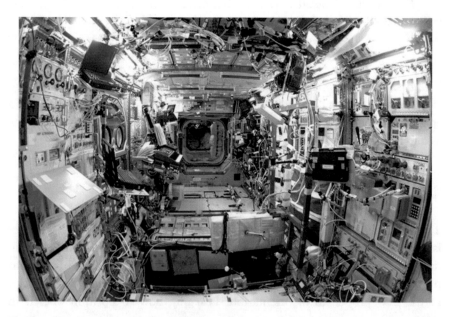

Fig. 4.1 The Destiny module of the ISS. Holding the book in different ways gives an impression of how the appearance of the room changes with orientation (Image: NASA)

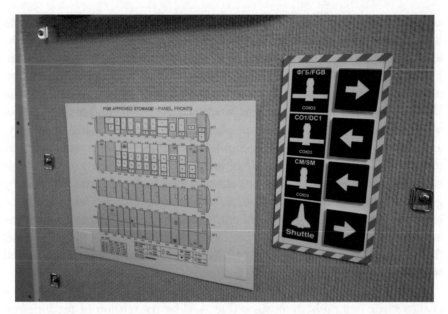

Fig. 4.2 Arrows showing what is where in the Russian segment of the ISS (Russian Orbital Segment, ROS). The picture was taken in the ISS mockup in the Johnson Space Center (Image: B. Ganse)

Fig. 4.3 Guided tour of the ISS by astronaut Sunita Williams. https://www.youtube.com/watch?v=doN4t5NKW-k

behaves differently and follows other rules than on Earth. Humans adapt so well to the new rules of weightlessness that it takes a while to re-adapt when back on the ground. Some spacefarers apparently had accidents when trying to walk through a doorway after return. Due to the changed behaviour of inertia and friction, they expected to arrive exactly in the doorway after touching the opposing wall with a finger! A coarse misjudgement! Instead, they hit the door frame. Another example of re-adaptation problems is that spacefarers tend to drop objects, since they got used to the fact that in space, things simply stay in place. These are typical examples of mistakes made during re-adaptation. Both anecdotes were told by German astronaut Hans Schlegel who had these problems himself. In addition, problems of coordination occur. The learning process for adaptation in the new environment takes about 1 month in weightlessness. The leg muscles are almost not required to produce force, while hands and fingers are used to push oneself off walls to gain momentum. ISS crew members sometimes got stuck standing still in the middle of a spacestation module (or deliberately placed there as a prank), without having any walls or objects within reach. Similar to the function of a rocket engine, they have to rely on the principle of recoil (see Fig. 2.11) to get out of this situation. One option is to throw a massive item in one direction to thereby get momentum in the opposite direction, another option is to generate an impulse by blowing air. This is an entirely different environment where the laws of physics appear to behave differently than on Earth. At the same time, fine motor skills are not affected at all. Writing and drawing work just as well (or badly) as on the ground.

In weightlessness, the human body will always instinctively return to a characteristic neutral body posture (Fig. 4.4) by the tension of tendons and muscles. This posture can especially be observed during sleep. Knees, hips and elbows are bent slightly, while the feet are stretched to 110°, and the head tilted forward by about 25°.

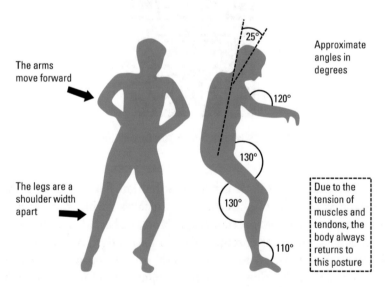

Fig. 4.4 The neutral body posture in weightlessness

Order and Mess

The untidiest flat, the messiest desk and the dirtiest junk room on Earth have one thing in common: the untidiness is still essentially two-dimensional! Items may form layers and chaotic piles or lie side-by-side, but gravity forces a certain vertical order on these items. In weightlessness, however, when even a few objects float around, the overwhelming concept of three-dimensional untidiness unveils its true nature. With no gravity present to hold things together, no self-stabilising vertical order appears on the scene. Instead, everything mixes up in an ununiform way, occupying more space than ever before and totally obscuring the view. Even the ISS, where its residents have taken good care to tidy up properly, always appears, depending on the definition, to be untidy. For this reason, it is important to never let items fly around in weightlessness. Instead, they should be attached, stored or fastened. A popular solution for this problem is velcro. Placed on most surfaces and objects, it allows to quickly attach and detach objects. Spacefarers often even have it on their pants. When designing a spacecraft, it is important to include enough storage space and velcro areas to prevent chaos!

> **Hint for Aspiring Spacefarers** If an item ever gets lost inside a spacecraft, such as a pen, a piece of paper or another object that could be flying around in the interior, it will probably be at the air inlet of the life support system! Since the air is constantly moved around by fans, the inlet grating is the most likely location for the air currents to transport it.

Thermoregulation and Airflow

Under the influence of gravity, warm air rises upwards and cool air drops down. When air warms up on the skin surface, it moves up while cooler air follows, and this cooler air then warms up on the skin. Through this process, a constant air exchange takes place, which is called convection. In weightlessness, however, this mechanism does not work (Fig. 4.5)! As there is neither up nor down, air stays where it is and forms a warm layer around the body that does not move unless altered by movement or airflow. The same is true for exhaled air, rich in carbon dioxide, that may form a bubble around the mouth and head. When the same air is inhaled and exhaled repeatedly, the decreasing oxygen and increasing carbon dioxide levels may put spacefarers at risk for intoxication. For this reason, a constant airflow is crucial on board spacecraft and spacestations!

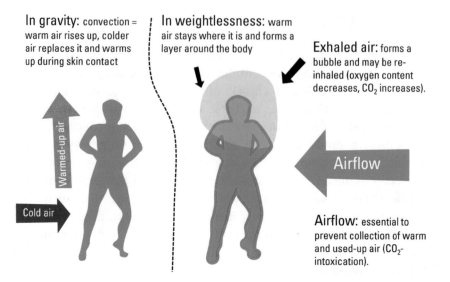

In gravity: convection = warm air rises up, colder air replaces it and warms up during skin contact

In weightlessness: warm air stays where it is and forms a layer around the body

Exhaled air: forms a bubble and may be re-inhaled (oxygen content decreases, CO_2 increases).

Warmed-up air

Cold air

Airflow

Airflow: essential to prevent collection of warm and used-up air (CO_2-intoxication).

Fig. 4.5 Principles of air circulation under the influence of gravity and in weightlessness

Humans have differing comfort zones when it comes to draft, just as with temperatures in general, that change with exercise and differ between people. This potentially minor issue may become a very sensitive topic. When buying a sleeping bag, the comfort temperature is usually labelled differently for the sexes. In space, a compromise needs to be found between the necessary airflow and comfort temperature. In case of problems, thicker clothing and warm drinks may be considered.

The Microbiome on Board

Growth of microscopic life forms such as bacteria and fungi on board space-craft and stations has received growing attention over the past years. Figure 4.6 shows black mould formation on the ISS. As mould is known for its potential negative health effects, such as respiratory tract infections, crews clean the International Space Station as part of their weekly schedule. Apart from its effects on humans, mould can also destroy surfaces, can change properties and even form holes in metal by biofilm formation. Filamentous fungi are abundant on Earth and show up in and on soil, structures, organisms, as well as in water. The ISS has been visited by hundreds of people, each of them bringing their own population of microorganisms along. These grow on all kinds of surfaces, similar to ground-based confined environments, such as intensive care units, operation rooms or submarines. The "EXTREMOPHILES"

Fig. 4.6 Mould formation on board the ISS on a panel where moist clothing and towels are usually kept to dry (Image: NASA)

experiment analysed the biome on board the ISS including swabs being taken in various locations. It showed fluctuations in the microbiome profile over time. Microbes are adapted to the environment, but not necessarily harmful to humans. For long-term spaceflight, microbial monitoring is desirable, and surfaces should be cleaned with potent disinfectants to prevent adverse health effects. While some fungi are resistant to radiation, many of them will disappear when exposed to UV-C light. Melanin-containing fungal species such as Aspergillus niger, on the other hand, resist UV radiation and cannot easily be removed with UV exposure alone. Evacuating the spacecraft by venting the atmosphere into space does not kill the microbiome reliably either, as exposure experiments on the outside of the ISS have shown fungal spores to survive for years in vacuum conditions. A promising approach to efficient decontamination of items in space vehicles are plasma processes with

Low Pressure Plasma (LPP). Other options include antibacterial surfaces, for example, made of copper or silver, that microbes do not feel comfortable on. A thorough overview of microbiological findings from the International Space Station is published by Mora et al. (2019).

4.2   Day and Night

The ISS orbits the Earth approximately every 92 min (see calculation in Sect. 3.3). During this time span, the crew experiences a sunrise, orbital day, sunset and orbital night. The length of day and night periods depends on the spacestation's orbit, especially the inclination against the equator. However, spacefarers cannot change their biological rhythm and sleep once every 90 min. Humans need much longer sleep periods to go through all characteristic sleep states necessary to recover. For this reason, a normal 24-h day is used for the schedule on spacestations. *Coordinated Universal Time* (UTC, identical to Greenwich Mean Time) is used on the ISS. Leap seconds are used to adjust UTC to the mean solar time to accommodate irregularities and slowing of the Earth's rotation. For shorter flights, *Mission Elapsed Time* (MET) is commonly used that shows the time since launch. In the early days of the Space Shuttle era, MET was still called *Ground Elapsed Time* (GET), but then changed to MET as it was often confused with Greenwich Mean Time (GMT). When planning a shorter spaceflight, MET is a more practical time, while for longer spaceflights and spacestations, UTC makes daily schedules easier to handle.

Circadian Rhythms

A *circadian rhythm* is the internal clock of living beings. It prepares the body for recurring events and, in mammals, is located in the suprachiasmatic nucleus of the brain. The scientific discipline that deals with these rhythms is called chronobiology. Many rhythms in humans have a period of 24 h, and several body functions and hormones fluctuate according to this cycle. These include body temperature, alertness and tiredness, blood pressure and the production of melatonin and cortisol. Disruptions of circadian rhythms lead to negative health effects, sleeping disorders and declines in performance. For this reason, good sleep has a very high priority in spaceflight. When travelling, synchronisation of the rhythms to a new time zone takes several days. The problems encountered during that period are called jetlag. Most circadian rhythms are not exactly 24 h long, but need timing sources, the so-called *zeitgeber*, such as short wavelength daylight, to synchronise. The pituitary gland in the brain secretes melatonin to regulate wake and sleep. Receptors in the eyes sense daylight and forward this information to the pituitary gland, thus making daylight the external timing source for the melatonin cycle. There are two extremes with regard to the preferred time of getting up in the morning (chronotypes): Eveningness (delayed, "night owls") and morningness (advanced, "early birds"). Many individuals find themselves somewhere in between or are able to discipline themselves and compensate on the weekends. Work schedules force many individuals to get up early and thereby act against their nature. Shift work with frequent changes between day and night shifts could lead to adverse health effects and sleep problems.

Concrete advice for space mission planning requires that a 24 h cycle should be established while in space, be it in a spacecraft or station. Eight hours of sleep should be scheduled so that they consistently occur at the same time interval in the cycle and not be divided or shifted. Artificial bright light may be used as a zeitgeber. Physical exercise should not be planned right before the sleep period. Incorporating daily rites to structure the day is very useful, such as having all meals together with the whole crew at fixed times, or reading before sleeping.

Sleep in Weightlessness

In theory, it is possible to sleep anywhere in any orientation, as long as the bed is fastened and does not float around freely. Floating might cause injuries and decrease sleep quality, as one would tend to wake up when hitting something. In space, most people like to sleep in a sleeping bag, preferably in a cosy and dark little compartment. The feeling of being hugged and having something touching the body from all sides is apparently beneficial for better sleep. Figure 4.7 shows the sleeping bag options of the Space Shuttle for comfy nights in space.

On the ISS, each crew member has their individual berth, a little room containing the personal items such as photos and notebooks, as well as their sleeping bag. Good air ventilation is crucial, as exhaled carbon dioxide otherwise collects to be re-breathed (compare Sect. 4.1). Sleep problems, sleeplessness and tiredness are significant problems in the daily life on spacestations and in spaceships. Many crew members take medication to sleep. Despite the recommendation to rest for 8 h, the average sleep duration is usually shorter than on Earth, comprising of only 6 h per day. This is caused by a combination of several reasons including missing the feeling of support from a mattress or bed underneath, light flashes from charged particles (space radiation), noise,

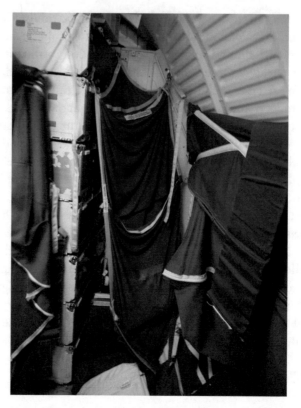

Fig. 4.7 Two different types of sleeping bags that were used during Space Shuttle flights. They are now displayed in the Shuttle mockup in Johnson Space Center (Image: B. Ganse)

increased stress levels and excitement. Problems may arise, for example, in the night before a spacewalk or other challenging task. Some spacefarers have reported intense dreams and increased snoring. It is a well-known fact that sleep quality correlates with performance. Good sleep is therefore the highest operational priority. Physical exercise, bright blue light during the day and warm light at night or melatonin are helpful. Melatonin can be consumed to improve the sleep quality and adjust the circadian rhythm.

When sleeping in space, many spacefarers report seeing light flashes and lines with their eyes closed. These are related to charged particles coming from the Sun or from the depth of space, which excite nerve impulses when hitting the retina. The impulses are transferred to the brain and processed in a way that may annoy or disturb sleep. On Earth, these light flashes do not exist, as particles are either deflected by the Earth's magnetic field or absorbed by the atmosphere. Since humans are not used to seeing radiation with their own eyes while shielded by the atmosphere, this is one of the more surprising effects of life in space. On other celestial bodies without a strong magnetic field or thick atmosphere, such as Mars or the Moon, similar light flashes would be experienced by humans during their stay. Section 5.7 will go into more detail of the physiological effects of radiation on humans.

For particularly long spaceflights, such as to the outer planets of the solar system or to other stars, a common idea from science fiction is to put people to sleep for the duration of travel and to wake them up in perfect health and shape on arrival. This artificial sleep-like condition is called *hypersleep* (if metabolic processes are halted completely) or *hibernation* (if they are slowed down). Ideally, the metabolism would be reduced to make travel without ageing possible and lessen resource consumption. As many mammals hibernate every winter, it should in theory be possible to create a similar state in humans, too. However, research in this area is progressing very slowly. Simply speaking, hibernation for humans is not currently possible or in reach.

4.3 Eating and Drinking

First of all: Yes, it is possible to swallow and eat normally in weightlessness, and no, it is not necessary to limit the culinary experience to puréed food in toothpaste tubes and liquids! Before the first human spaceflight, great uncertainty existed concerning this matter. Luckily, it quickly became apparent that eating and drinking function without problems in space. It was, though, rather complicated to find ways to prepare food in a way that is practical to handle in zero gravity and easy to prepare on board. While food was contained in toothpaste tubes in the early 1960s, progress was made quickly. Menu options increased in numbers and a wider variety of food became available. In the Apollo missions, the availability of hot water allowed for more options and freeze-dried meals became common. The American spacestation Skylab in the 1970s contained a fridge and a freezer, and a selection of 72 different menu items were available. In the space programme of the Soviet Union, food in tubes and cans were used for a much longer period, offering delicacies such as borscht, tongue and caviar. Today, many food options are available, and besides ready-made and freeze-dried meals, fresh vegetables and fruit are delivered by arriving spaceships and transporters. Salads and other vegetables can be grown in space too (see Sect. 6.5), but are currently mainly used for experimental purposes rather than to seriously and substantially provide the crew's nutrition. Several companies have developed food for spaceflight consumption, and some of it is quite normally sold in supermarkets and outdoor stores. Figure 4.10 shows food and packaging from the ISS. Russian cosmonauts currently have a selection of more than 300 meals available. Also Chinese, Japanese and Korean food has been consumed in space. Spacefarers usually participate in a *foodtasting* before the flight, where they can try and choose food. One should not underestimate the amount of change in taste perception in weightlessness! On a normal day on a spacestation, humans need approximately the same amount of calories as on Earth, as the basic metabolic rate remains the same. Several formulas are available to compute the caloric requirement. It depends on factors such as sex, muscle mass, body weight and age. Many more calories, however, are consumed during spacewalks, as these are usually very strenuous and exhausting (Figs. 4.8 and 4.9).

Hint for Aspiring Spacefarers When making coffee on Earth, hot water runs through the coffee powder and a filter. This process, however, requires gravity! So how can coffee be made in weightlessness?

In 2007, the ISS expedition 7 invented a simple method: First, coffee powder and hot water are mixed in a cup. Next, a filter or sieve is put on top of the liquid in the cup. The cup is then sealed by a lid and … stuck into a sock! In the sock, the cup can now be spun around at high speed, generating centrifugal forces that push the filter towards the bottom of the cup. This way, the liquid separates from the powder, and the result is reportedly delicious coffee, which can be drunk with a straw.

In 2014, this method became obsolete, when the Italian astronaut Samantha Cristoforetti installed the new coffee maker *ISSpresso* to brew the first espresso in space.

To drink the coffee, an alternative to a straw is available in the form of special zero gravity cups that prevent their content from floating around freely. Their rim is not circular, but they have a pointy mouthpiece that keeps the fluid attached by capillary forces. US astronaut Don Pettit built the first cup of this kind while goofing around with a plastic sheet from a folder.

Anecdote

In 1961, during the first human spaceflight (duration: 1 h 48 min), cosmonaut Yuri Gagarin ate the content of three toothpaste tubes of 160 g each: two containing puréed meat and one filled with chocolate sauce.

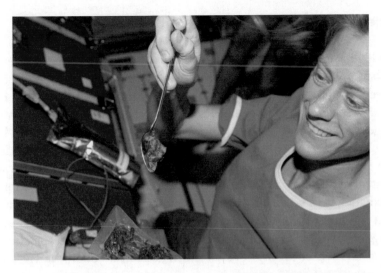

Fig. 4.8 NASA astronaut Karen Nyberg enjoys a meal of creamed spinach in weight-lessness (Image: NASA)

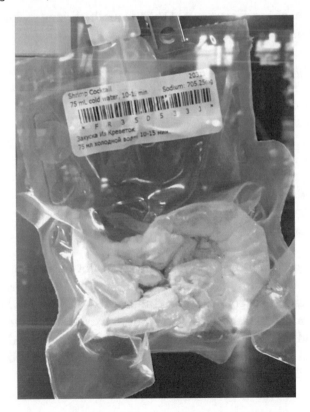

Fig. 4.9 Shrimp cocktail in its ISS packaging. Not the most appealing meal to look at, but apparently the most delicious one in space (Image: B. Ganse)

Sense of Taste Upside Down

Already in the early years of human spaceflight, it became apparent that taste and food preference change drastically in space. Taste buds seem to be altered by swelling due to the fluid shift in weightlessness, similar to when having a common cold. Spacefarers therefore usually prefer more intense tastes and stronger spices. This is, however, not true for everyone in the same way. Dishes that fulfil a culinary niche at best on Earth are the stars in the space kitchen, above all the *shrimp cocktail*. It mainly consists of shrimp, sauce and spices. This item of astronaut food has been available since the Gemini missions and immediately became the most requested food in space. Even astronauts who would not fancy it in their own kitchen, or found it disgusting in the food tastings on Earth, consider it to be the most delicious dish available. Rumours say that servings of shrimp cocktail get traded like an unofficial currency on the International Space Station.

Hint for Aspiring Spacefarers Do it yourself shrimp cocktail (typical Earth recipe)! Ingredients:

- 350 g big shrimps
- 4 Tbsp. mayonnaise
- 2 Tbsp. cream
- 1 Tbsp. sherry
- 2 Tbsp. ketchup, cut tomatoes may be added
- 1 lemon or some lemon juice
- Salad
- Onion flakes
- Tabasco, horseradish, garlic or Worcestershire sauce
- Celery salt and pepper

Shrimp cocktail is best served in a glass, or for the proper spaceflight experience in a plastic pouch. To make the sauce, mayonnaise, cream sherry, ketchup and lemon are mixed and stirred. Salt, pepper, liquids and spices are added to taste. The salad is put in the bottom of the glass. The shrimp and sauce are added on top. Shrimps may be arranged in a circle or added to the sauce. A slice of lemon or tomato may be added to garnish the dish. Enjoy your meal!

Packing and Preparing Dishes

Food on board a spacecraft or station needs to fulfil several requirements: it needs to be as lightweight as possible, provide energy, be delicious, quickly prepared, easy to digest and should not cause problems in weightlessness. This means it should stick together and not end up as crumbs or droplets in the air filters. Carbonated drinks may cause discomfort and pain when air bubbles collect in the bowels and stomach, as they spread out locally rather than moving up. Packaging needs to be light and easy to use. Vacuum sealed plastic and foil-laminated pouches (Fig. 4.10), as well as cans seem to work well. Fresh fruit and vegetables should be eaten within a couple of days. Nuts and cereal bars can be easily packed and stored for a longer time. Meat is usually treated with radiation to prevent growth of bacteria and fungi. All food items need to come readily prepared as frying or roasting is not possible in space. Heating is possible in a small forced convection oven on the ISS, which can be used to bake a single cookie at a time. For some dishes, water needs to be added before consumption. Those have previously been treated with either heat, freeze-dried or osmotic conditioning.

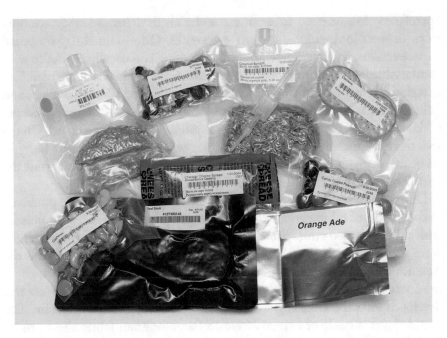

Fig. 4.10 Examples of packaged food and drink items that are available on the ISS (Image: NASA)

Hint for Aspiring Spacefarers Many outdoor stores sell food for hiking and trekking that has been prepared in the same way as food on board spaceships and spacestations. Some of them may even be identical to the ones served in space (the packaging might not even mention this fact)!

Anecdote

For the Gemini 3 mission, astronaut John Young smuggled a corned beef sandwich on board the spaceship. During the flight, he shared it with colleague Gus Grissom, but they did not finish the meal as it started to crumble in the cabin. When NASA officials heard about this misconduct, they were not amused, as crumbs could have interfered with the electronics on board and endangered the mission. Young was temporarily suspended from active flight duty, but was eventually readmitted.

4.4   Clothing

Spacefarers do not need to wear spacesuits all day! Instead, they are only worn in specific situations, such as spacewalks. On a normal day on board a spacestation, the crew can wear normal clothing and usually choose to wear comfortable items. Fabrics that do not cause strong body odour are useful, so that clothing may be worn longer without being washed. T-Shirts, sweaters and polo shirts are common on spacestations, while clothing with long appendages or dangling rope loops would not meet the safety regulations on board. On the ISS, there is currently no washing machine, which is why clothes are brought on board, worn and then discarded. For future long-term missions outside the Earth's proximity, the ability to wash will be essential.

During launch and landing, spacefarers wear *pressure suits* to be protected in case of loss of cabin pressure. These suits can provide sufficient air pressure with a 100% oxygen atmosphere, and they are less complex than the spacesuits worn during spacewalks (see Sect. 4.5). Examples of pressure suits are the American *ACES* (Advanced Crew Escape Suit, Fig. 4.11) that was used in the Space Shuttle and the Russian *Sokol* suit (Fig. 5.10). The ACES suit was developed after the Challenger disaster to better protect the astronauts. The outside layer is made of orange *Nomex*, a flame-resistant fabric also worn by firefighters. Under it, an airtight layer acts as pressure vessel just like in a dry suit for divers. After the end of the Space Shuttle programme, NASA had

Fig. 4.11 The American astronaut and medical doctor Dr. Mae Jemison in her ACES suit during a last check before launch of STS-47 in 1992 (a Spacelab mission). She was the first Afro-American woman in space and has published an exciting autobiography *Find Where The Wind Goes* (Jemison 2003) (Image: NASA)

initially announced not to continue the ACES suit. However, for the new Orion spacecraft, they have now developed it further to be useable for both launch/landing and spacewalks. The new suit is called *OCSS* (Orion Crew Survival System). Compared to ACES, OCSS offers a higher range of motion and a *closed loop system* in which breathing gases are regenerated.

The Sokol suit has both an outer and inner layer of Nylon, a synthetic polymer. Airtightness is achieved by the rubberised, thick inner layer. The gloves of the suit attach via metal locking rings made from anodised aluminium. The visor is made of polycarbonate. The rest of the helmet is non-rigid material, so that it can be collapsed when the visor is open. China has bought Russian Sokol suits for their own space programme and copied the style to have a very similar suit.

4.5   Spacewalks

In technical jargon, spacewalks are called *extravehicular activities* (EVA). Per definition, the term includes all human activities outside a spacecraft, be it in an orbit or on the surface of a celestial body. Hence, the 28 walks that were performed on the Moon are also referred to as EVAs. For any EVA, a spacesuit is required, as the Earth is the only (known) place with an atmosphere breathable for humans. Spacesuits serve as temporary one-person spacecraft, as their life support systems enable humans to survive in an otherwise deadly environment. Alexei Leonov conducted the first EVA in 1965 during the Voskhod 2 mission. The EVA took 12 min and almost ended in a disaster, when his spacesuit inflated way more than expected and the hatch was too small for him to fit back in. Leonov had to release air from his spacesuit through a valve to be able to reenter the spaceship. The loss in pressure led him to experience symptoms of decompression sickness (muscle and joint pain from bubbles forming in the tissue, see Sect. 5.5). When he had finally made it back inside, the hatch did not close properly and the life support system started pumping oxygen into the cabin to compensate for the loss of pressure. These high oxygen levels implied a significant fire risk, since the spaceship was not designed for oxygen concentrations this high. Luckily, Leonov and his colleague Pavel Belyayev were able to complete the mission as planned. The first EVA for the US was conducted shortly after by Ed White. Since then, several hundred EVAs have been completed successfully. China, represented by taikonaut Zhai Zhigang, has become the third nation to carry out an EVA with its own spacesuit technology during the Shenzhou 7 mission in 2008.

EVAs have, however, been associated with problems and surprises time and time again. They should still be regarded as a risky business. Problems include suit leakage, frostbite, blood collecting under the fingernails (haematoma) and impaired vision due to water condensation. In 2013, Italian astronaut Luca Parmitano's EVA was aborted early, when water from the spacesuits cooling system started to collect in the helmet and slowly moved forward to his nose and mouth due to a technical problem of the suit. On the way back to the airlock, the water covered his eyes and went into his nostrils. He could not see clearly through the water and was only able to breathe through his mouth. His microphone stopped working, too. With the help of his buddy[1] Chris Cassidy, he was able to reach the airlock in time. To reduce risks, many regulations apply to EVAs and the procedures are clearly defined and rehearsed many times. In

---

[1]"Buddy" is the official terminology here, similar to the diving buddy.

case of an emergency that makes a spacewalker unable to react, colleagues need to bring him or her to the airlock.

While working on the Moon during the Apollo programme, astronauts had serious problems with lunar dust. It consists of very small, gritty particles of regolith that are very hard to remove and stick to any surface, causing abrasion. Apart from damaging technology, it might also have negative health effects on the crew. Similar problems have been suggested for Martian dust (see Sect. 6.4). During the Apollo programme, NASA developed *Moon Boots*, which are special shoe covers to protect from the lunar dust (Fig. 4.12).

> **Hint for Aspiring Spacefarers** When exploring a new planet for the first time, as a precautionary measure, the spacesuit should be left on and the helmet visor should remain closed, even if instruments show sufficient pressure and oxygen content in the atmosphere. Dust might be toxic and carry other nasty chemicals with it. And who knows what biological surprises might be found?

Details of Spacesuits for EVAs

Spacesuits for EVAs, other than pressure suits worn inside spaceships, need to be little spaceships themselves. They are required to provide breathable air with sufficient pressure and oxygen content for several hours, regulate the temperature, provide drinking water and protect against micrometeoroids. NASA's

Fig. 4.12 Apollo *Moon Boots* from 1969 are shoe covers to protect the lunar lander's inside from moondust (Image: B. Ganse)

EVA suits are called "Extravehicular Mobility Units" (EMUs, Fig. 4.13). The Russian counterpart is the "Orlan" suit, the Chinese equivalent is called "Feitian" (Mandarin "flying sky"). The Orlan suit was initially developed for the Soviet Moon programme. Even though no Soviet ever made it to the Moon, the suit was developed further. Currently, the fifth version, Orlan-MK, is used on the ISS. The crucial part of an EVA suit is its life support system that regulates the pressure, gas concentrations and temperature of the air. It also removes carbon dioxide (see Sect. 2.5). Another important function is urine collection to prevent smells and discomfort. In addition,

Fig. 4.13 Scott Kelly in the *Extravehicular Mobility Unit* (EMU) 2015 (Image: NASA)

systems for communication and telemetry (monitoring of the heart rate and further important parameters) are provided. In both types, EMU and Orlan, the helmet is fixed to the torso, which means the head is turned within the helmet to look sideways through a large visor. During an EVA, air pressure is reduced to 30 kPa (40 kPa for the Russian Orlan suit), which is only a third of the usual pressure on board a space station. Reasons include that it is much easier to move when the suit is inflated with less pressure. Before an EVA, slow adaptation to lower pressures is required to prevent decompression sickness. Astronauts in the EMU breathe pure oxygen without nitrogen. The EMU allows EVAs with a duration of 8.5 h, while 9 h are possible in the Orlan. Spacesuits often have many connectors and hoses. Figure 4.14 uses the example of an Apollo spacesuit to explain the purpose of each connector.

In the 1970s, a very special spacesuit was developed that allowed for untethered independent propulsion, the *Manned Maneuvering Unit* (MMU, Fig. 4.15). Individuals could independently manoeuvre and manually steer it to fly around in space with it. The unit was used during three Space Shuttle missions in 1984 and then retired, as safety concerns arose. In theory, an astronaut could have run out of propellant and been lost in space without a possibility to be rescued. This was not unlikely to happen, since orbital mechanics are non-intuitive, as described in Sect. 3.3. The successor model SAFER (*Simplified Aid for EVA Rescue*) is an emergency safety system attached to each EMU backpack. Cold gas thrusters using gaseous nitrogen are propelling both models, each SAFER having 1.4 kg propellant on board.

How to Perform a Spacewalk

Every single step and hand movement of a spacewalk is practised intensively before the flight in a large, deep swimming pool with an underwater spacestation mockup. Some of these pools are much larger and deeper than an average Olympic swimming pool (with a depth of less than 5 m). In addition, they host an array of extra equipment such as underwater cameras, cranes and communication systems. NASA's pool is located in Houston and called the *Natural Buoyancy Laboratory* (Figs. 4.16 and 4.17). Its dimensions are impressive: it is 62 m in length, 31 m wide and 12.34 m deep. ESA has a smaller pool, called the *ESA Neutral Buoyancy Facility* in the *European Astronaut Centre* in Cologne, Germany. Its size is 22 × 17 m, with a depth of 10 m. Russia's circular *Hydro Lab* has a diameter of 23 m and a depth of 12 m. The Chinese NBF is located at the *China Astronaut Research and Training Center* in Beijing (size: 23 m in diameter and 10 m in depth). The Japanese circular *Weightlessness*

Fig. 4.14 The spacesuit worn by Alan Shepard on the Moon in Apollo 14. The three left connectors are: top: communication, middle: external air inlet, bottom: external air outlet. On the right: top: water inlet, middle: backpack air inlet, bottom: backpack air outlet. For more details see the NASA-handbook *Extravehicular Mobility Unit* (Image: B. Ganse)

Environment Test System (WETS) facility was severely damaged and closed in a 2011 earthquake. Astronauts have reported differences to a real spaceflight, since water resistance is missing in space; yet on the Earth, training in the pool seems to be the best way to practise EVAs.

To reduce the risk of decompression during an EVA, it is necessary to prepare in a specific way. Spontaneous EVAs without preparation are irresponsible and dangerous! In case of a leak in the spacesuit, be it caused by collision with an object or damage to the suit, the sudden drop in air pressure can again cause symptoms of decompression sickness. To prevent this, as much

Fig. 4.15   First EVA using the *Manned Maneuvering Unit* in 1983 by Bruce McCandless without a connection to the Space Shuttle, shown from a distance of more than 100 m (Image: NASA)

nitrogen as possible needs to be removed from the body's tissues before each EVA. One option is to breathe pure oxygen for several hours prior to the EVA (*prebreathing*). This method works and is the standard process in crewed spaceflight. The protocol is called "In-Suit Light Exercise (ISLE) Prebreathe Protocol". However, an alternative procedure was previously used on board the ISS since 2006. It is called *camping out*. To remove the nitrogen from their bodies, spacewalkers sleep in the airlock with 100% oxygen during the night before their EVA. The airlock used for this purpose on the ISS is called *Quest* and can be applied with both American and Russian EVA suits. During the night, the pressure is slowly reduced to one-third. In the morning, the spacefarers are allowed to briefly visit the bathroom and have breakfast with an oxygen mask before putting on their spacesuits and starting their EVA. The hatch can be opened once all air has been pumped away from the airlock,

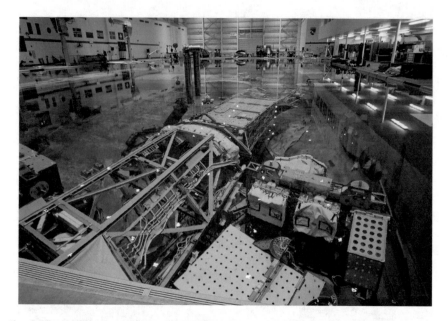

Fig. 4.16 NASA's *Natural Buoyancy Laboratory* in Houston (Image: B. Ganse)

Fig. 4.17 NASA astronaut Peggy Whitson trains in the *Natural Buoyancy Laboratory* (Image: NASA)

so that no pressure difference to the vacuum outside remains. After return from the spacewalk, the pressure in the airlock is slowly elevated again over several minutes before entry to the station is possible. Pressure changes need to occur slowly and according to specific rules just like in diving, to avoid decompression sickness.

4.6 Research

The two main tasks of spacefarers on board spacestations are usually to conduct research and to be guinea pigs. Scientists propose and train the crews on their experiments that will be conducted by the crew on board. Space agencies regularly solicit *announcements of opportunity* for researchers to submit their suggestions for projects. Find more details on how this works in Sect. 5.9.

Spacefarers participate in research experiments as subjects. They choose from a variety of options. Each individual is supposed to conduct as many experiments as possible, but will not be forced to do something they do not feel comfortable with. The scientists who suggest the studies write information sheets on their experiments, and spacefarers need to sign consent forms, just like all participants of human studies on Earth. In Europe and in the US, astronaut trainers rehearse the experiments and practise them with the spacefarers before the flight. Here are some examples of the many experiments conducted by spacefarers on the ISS.

The EXPOSE facility is a box that can be mounted to the outside of the ISS in a spacewalk to see how biological and chemical samples or other materials react to exposure to the open space environment (Fig. 4.18). Astrobiologists are particularly interested in how far bacteria and other microscopic life can survive radiation, extreme temperatures and the vacuum of space. One of the fundamental questions is whether life might have been brought to the Earth by asteroids from other sources of origin. Other questions include under which conditions life is able to survive and replicate. Samples are kept in little chambers within the EXPOSE facility. Each chamber is covered by a number of possible optical filters to regulate UV light spectrum and intensity (which also prevent loss of the sample). As an alternative to vacuum within the chambers, it is also possible to simulate the atmospheres of other planets, such as Mars. Some samples have been exposed for over 1 year. At the same time, doses of several different types of radiation are measured with dosimeters. An example of an experiment that has been run on the EXPOSE facility is PROTECT. The experiment PROTECT measured the survival of the bacterium Bacillus subtilis and B. pumilus in conditions corresponding to a

Fig. 4.18 Cosmonaut Dmitri Kondratyev inspects the EXPOSE facility to send it back to the ground for analysis (Image: NASA)

travel to Mars and presence in the Martian atmosphere. Results indicate that DNA and protein damage were caused by exposure to solar UV radiation in space and in the atmosphere of Mars, further aggravated by the vacuum of space. Another result was that spores survived much better when arranged in several layers, compared to only one layer.

Another device used for science is the *Microgravity Science Glovebox* that allows experiments with (hazardous) substances and prevents them from floating around freely in the spacestation. It also allows for combustion experiments. Working in the box also inhibits contamination of samples. Reduced pressure inside the box can be used for specific experiments, and an airlock helps bring items inside. The work volume is $91 \times 61 \times 46$ cm. The operator can manipulate objects inside the closed system using rubber gloves (Fig. 4.19). An example of an experiment performed inside the glovebox is the PromISS-2 experiment that studied growth processes of protein crystals in microgravity.

Medical research is often carried out by spacefarers on one another or on themselves. They usually learn to take their own blood from one arm using the other arm. Blood samples can be centrifuged and stored in a $-80\,°C$ freezer. Samples are brought back to the Earth with either uncrewed transporters or

Fig. 4.19 American-British astronaut Michael Foale checks the *Microgravity Science Glovebox* (Image: NASA)

spacecraft together with the crew. Human research can even be conducted during sleep or exercise. Many medical diagnostic devices are on board the ISS. One of the authors is involved in an ISS experiment testing the efficacy of electrical muscle stimulation in addition to the exercise on-board. The spacefarers are required to apply the electrical stimulation device by themselves and take and store their own blood samples.

4.7   Emergencies on Board

Spacecraft are among the most intricate machines humans built, and therefore many things can and will go wrong. A lot of information on medical emergencies on board will be provided in Chap. 5. Therefore, the present section is limited to non-medical incidents and their management. First of all, the response to major emergencies depends on whether or not immediate return to Earth is possible. From the ISS, reentry within a few hours can be achieved at any time. On long-term missions to other planets, moons and asteroids, flying home will not be an immediate option. Possible emergencies include:

- Sudden loss of cabin pressure, for example, when hit by a micrometeoroid,
- Leakage of a poisonous substance or gas such as ammonia from the cooling system, or hydrazine from a propellant tank,
- Failure of the life support system,
- An explosion on board, for example, of an oxygen tank,
- Loss of attitude control or propulsion due to computer or engine failure.

Clearly defined procedures are required that state what to do in case of an emergency. Crew escape systems are often installed to enable escape from spaceships (see Sect. 3.1 and Fig. 4.20). In addition, duplicates of important systems and alternative options are necessary (Sect. 2.5). Emergency procedures for the ISS are documented in the file linked by QR-code in Fig. 4.21. The document explicitly defines the steps for a variety of situations. As an example, in case of a leak in the international part of the ISS, everyone is asked to move to the Russian Zvezda-module to analyse the problem further from there. As soon as the exact location of the incident is known, the

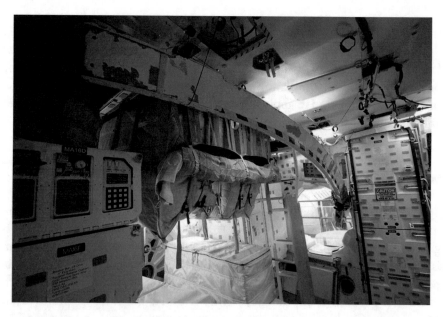

Fig. 4.20 The Space Shuttle's Crew Escape System was a beam to extend from the Shuttle. The crew could use it to escape in case of controlled gliding when unable to land on a runway (Image: B. Ganse)

Fig. 4.21 Document containing the official emergency procedures of the International Space Station http://www.spaceref.com/iss/ops/iss.emergency.ops.pdf

particular module shall be isolated and a repair procedure can be planned. If possible, it is important to re-establish the atmosphere in the module. In case fire extinguishers are used, one needs to be aware of their acceleration effect in weightlessness and hold on to something to not float away! The air can be investigated using a CSA-CP (*Compound Specific Analyzer—Combustion Products*) and other diagnostic devices to find residues like carbon monoxide or other toxic gases. The ISS's life support systems have been built in a redundant way (there are a Russian, an American and a European system), so that survival is possible even if part of the station has been isolated. Evacuation of the entire station would only be necessary in the most serious of events and has luckily never happened.

Examples of on-board emergencies in spaceflight history include the explosion on board Apollo 13, the fire on board Mir spacestation and a hole in the hull of a Soyuz spacecraft. During the Apollo 13 mission, the explosion of a liquid oxygen tank led to a loss of power in the Service Module and put the mission and its crew at great risk. Only ingenious and creative on-board engineering and the use of the Lunar Module for life support could rescue the three men. In 1997, a lithium perchlorate canister used to generate oxygen burned on the spacestation Mir. The fire was put out with a fire extinguisher after 90 s, but smoke needed several minutes to clear. The crew had to wear oxygen masks and follow safety procedures. In 2018, a dip in cabin pressure led to the detection of a hole of 2 mm in the Soyuz MS-09 spaceship. It was patched up with epoxy and did not cause further problems. It later turned out that a ground technician had accidentally drilled this hole and mended it with super glue to hide his blunder.

So far, luckily no fatality has been reported in Earth orbit and beyond, except for during launch and landing. In case a crew member passes away, opening an airlock and throwing the body out of the spaceship is not advisable.

The body will likewise remain in orbit and may collide with another object. Airtight sacks for storage and retrieval should be available on any spacecraft. Hopefully they will never be needed!

To protect the general population from the dangers of spacecraft accidents, self-destruct systems (similar to the ones in the Star Trek movies) are built into many rockets and can be triggered by a Range Safety Officer (see Sect. 3.1). Historically, such a mechanism has only been triggered once in a crewed flight, when the Space Shuttle Challenger broke up shortly after launch, and its solid rocket boosters continued to fly onwards uncontrollably. The boosters were blown up by a radio command from the ground.

4.8   Social Life in Isolation

Social life and psychological aspects are closely allied in spaceflight and of high importance for mission success. Long-term spaceflight and human space exploration require prolonged stays in isolated and confined environments. They are associated with real risks and threats to the crew's health and life. Intactness and functioning of each crew member individually, as well as in the team, called the human factor, is the key for a successful completion of the mission. Many human isolation studies and space missions have demonstrated a variety of psychosocial problems associated with long-term isolation with limited resources. These include social stress and conflicts among the crew, changes in emotional state, team psychology, crew cohesion and group dynamics. These problems and conflicts have been known since the early days of human spaceflight.

Ground-based analogue isolation research studies have a history over several centuries in studying and improving the understanding of conflicts and the psychology as well as physiology of isolation and confinement. In the 2 year long closure experiment of Biosphere 2, conducted between 1991 and 1993, the eight participants showed a variety of problems, including massive conflicts, development of factions and disagreements on overall mission procedures. Apart from social and psychological effects of isolation and confinement, the Mars 500-experiment also showed substantial physiological changes developing over the confinement period of 520 days. The reason is that participants became increasingly inactive with a reduction in the variety of exercises throughout the isolation period. Also during a 30-day isolation experiment on board the *Ben Franklin (PX-15)* mesoscaphe (a type of submarine) in 1969, tensions became the most critical issue for mission success.

Communication and the understanding of what is going on play important roles in conflict management. Conflicts can best be solved when they are detected early and an intervention is applied before escalation occurs. As these issues play central roles for a successful spaceflight, space agencies are currently investing millions on research in this field.

How to Deal with Each Other

Frequent causes for tensions in a group are a lack in psychological compatibility, insufficient leadership skills, undefined roles, unavailable private space and cultural differences (for example, background, habits, religion or smells). Some behaviours are known to be rather problematic in a group and need to be stopped immediately when observed. These include scapegoating, which is a frequent behaviour in groups that leads to a decreased group cohesion. Scapegoating is present when a person is made responsible for things that went wrong, despite being innocent and often without even being involved. Other negative developments are crew members becoming outsiders or formation of subgroups among the crew. These behaviours may be reduced by specific social training and leadership courses. Everyone needs to understand the potential consequences for the group and act accordingly. Practising leadership is most crucial. In social psychology, according to Kurt Lewin, three styles of leadership exist: the authoritarian, the democratic and the laissez-faire style (Lewin et al. 1939). The authoritarian style is defined by hierarchy, while the democratic level is characterised by community decisions. When leading in a laissez-faire way, a lot of freedom and independence are given to each team member. Even though the concept to separately evaluate these styles of leadership in research is considered outdated, for most team members, it makes a big difference if a more military or a cooperative/democratic leadership culture is followed. The person in the leadership role should be able to show skill to find the right way to deal and communicate with each team member, motivate and integrate them. Many research studies have been conducted on topics of psychosocial aspects of spaceflight, and the results indicate that the commander has a key function for group cohesion and needs to be trained in actively managing it.

In the past, tension and negative emotions have often caused problems between crew and *Mission Control*, which is called "Us versus Them" syndrome. The pattern is as follows: The crew comes to the conclusion that the team at home is clueless and does not know the reality of spaceflight and isolation. They start to increasingly believe that Ground Control is no help and will not be able to give substantial support if really needed and are jealous of their good food, sunshine and all the pleasures of life. The crew feels left alone in regard to threats or dangers and perceives the team at home as superfluous. If Ground Control now starts telling the crew what to do and maybe even dictates a tight and demanding schedule, conflicts may appear. In order to prevent this from happening, it is important to involve close friends with similar previous experiences to communicate between the crew and the team on the ground. It is best to choose experienced spacefarers who are well known and trustworthy to the crew. They will allow for better communication and reduce the likelihood for conflicts in the frame of an "Us versus Them" situation.

Anecdote

In 1973, the third crew on board spacestation *Skylab* complained of heavy workloads that they were hardly able to cope with. All of them were in space for the first time. When *Mission Control* repeatedly ignored their complaints, the crew decided to take a day off. They turned off radio communication and enjoyed a relaxing free day without experiments and work on board. In the meantime, Mission Control had the hardest time on the ground in great concern and anger. When the crew turned the radio back on again, the ground team was more than happy to reduce the workload and to respond to the crew's needs.

Interpersonal problems increase by 20% in the second half of a mission. Also the variety of topics the crew members talk about decreases throughout a mission. At the same time the group starts to filter information they share with the outside world. This behaviour is called *Psychological Closing* (Gushin et al. 1997). It can be observed in many areas of daily life, too. Details about activities or conflicts may be dealt with by the group, without mentioning them to outsiders. Another typical behaviour that most people have already come across in their daily life is that references or jokes are shared by the team but at the same time not explained to others. This is to demonstrate their strength as a group and make the others feel separated. Arrival of new people on board a spacestation, as well as the departure of crew, are psychological burdens. It takes effort to get used to each other, find a new social order, habits of what to communicate with whom and learn not to get on each other's nerves. The unplanned elongation of a stay in space is usually perceived as very difficult. During ISS Expedition 43 in 2015, an uncrewed Progress spacecraft exploded, causing a debris field. The debris field delayed the return of the crew by 4 weeks. These individuals found it particularly hard to stay longer, as they had been looking forward to seeing their family and friends so badly. It is in general especially hard to miss out on important family events. Crew members should be given the chance to talk to their family as much as possible if such a situation occurs. Another situation hard to cope with is when family members die on Earth. Italian astronaut Paolo Nespoli was on board the ISS when his mother passed away. He was unable to attend the funeral and meet his family.

Group cohesion improves the more time a team has shared together before the mission, and this effect should be recommended in spaceflight. In science-fiction movies, the crew sometimes meets for the first time in-flight on board the ship (the 2014 movie Interstellar is an example for this). It is a good idea to give crew members as much control over their time schedule as possible and to avoid overwork. In addition, psychological counselling during the flight shows positive effects. However, counselling gets more difficult with increasing distance to the Earth when a proper conversation with someone at home is not possible due to the lag in communication, and messages can only be exchanged with delays. Communication with family and friends should be scheduled in regular intervals to have something to look forward to. Furthermore, studies have shown that teams function better when consisting of mixed genders. Stereotypes and unconscious prejudices against other nationalities, sexes and backgrounds should already be addressed early in the training and clarified beforehand.

Psychological Stability

In spaceflight, psychological stability depends on many factors. Real threads to life, confinement, boredom, monotony, isolation from family and friends and conflicts with other crew members may have a negative influence on psychological well-being. During longer missions, some individuals may develop a state of seclusion, exhaustion, loss of interest and motivation, irritability and potentially psychosomatic symptoms such as pain. This state mainly shows up in the third quarter of a long mission and is therefore called the *Third-Quarter-Phenomenon*. In long-term missions (with a time span longer than half a year), the initial excitement of the adventure slowly declines into relative boredom and may progress to psychological problems in some individuals. These may include anxiety, aggression and symptoms comparable to depression. Towards the end of the mission, euphoria suddenly arises and largely improves the situation. The crew now looks forward to coming home. Psychological evaluation of candidates plays an important role and is able to detect likely more vulnerable individuals. Testing includes questionnaires and interviews. Important factors are mental stability, the lack of mood swings or aggressions, a plausible and real interest in the mission, excitement for participation, social competence and a controlled, self-aware behaviour. Spacefarers get a lot of media attention and often have to deal with their role as celebrities. This includes being recognised on the street, having fans, being a role model and being present in the media despite having a bad day. Media attention can be very challenging.

Leisure Time

And now to a very nice topic: How to spend free time on a space mission in isolation and confinement? Recreation plays an important role for the well-being and psychological stability of spacefarers. Monotony and boredom absolutely need to be avoided. During the very first, short spaceflights of human history, every second was filled with work and survival had the primary attention. The issue of leisure time arose when people started to spend longer periods on spacestations. On the ISS, free time is used to relax, communicate over the internet and to share experiences with the world. Interplanetary spaceflights will most likely involve lots of spare time that is not consumed by routine work, training or experiments. From a psychological point of view, boredom should be avoided. First of all, it is important to spend time with the other crew members and celebrate daily routines together. Birthday parties and celebrations of all sorts of events are very beneficial. A morning meeting is recommended as well to go through the duties of the day. Shared meals and planned contact to family and friends are further important items. Due to the increasing lag in communication, it will only be possible to have a proper conversation with someone on the Earth and to play online games while close to the home planet, but this option soon vanishes. The internet protocol TCP/IP has a built-in time-out after 90 s. Therefore, the signal only has a maximum of 90 s round trip from the spaceship to the Earth and back. If the data packages take longer, they are considered lost. This means, the normal internet stops working in a distance of 45 light seconds from Earth, which corresponds to approximately 13 million kilometres. In comparison, the closest distance between Earth and Mars in the last years was 56 million kilometres in 2003. When outside of the TCP/IP range, other forms of data packages may be sent to communicate, but surfing the web is not available as a leisure time enjoyment. Activities such as exercise, computer and other games, reading, writing, drawing, photography and virtual reality are, among others, possible options to spend time.

Valuable experience for long-term space travel and potential future colonies have been collected in the Arctic and Antarctic. Here, humans are usually isolated in a hostile environment for many months. Several countries have research stations of varying sizes in both the Arctic and Antarctic. Space agencies conduct research and offer research opportunities to scientists on a regular basis. Many countries run mining towns in the Arctic, where many people live their daily lives surrounded by glaciers and permafrost. An example for such a settlement is the now abandoned Russian *Pyramiden*, located on the Svalbard archipelago in the Arctic at 79° north (Fig. 4.22).

More than 1000 people used to live here, including workers and their families with children. The settlement was founded in 1910 and abandoned in 1998. Each winter, the place is covered in darkness of the polar night for 3 months. Pyramiden, however, is a brilliant example for successful psychosocial integration in isolation. The settlement was equipped with a number of facilities to allow for an intact social life with a variety of facilities for activities. These included a big sports centre with a large indoor swimming pool, gyms, a cinema, choirs and orchestras, and many more. The place also had excellent facilities including a hospital, school and kindergarten. Animal stables provided fresh eggs, milk and meat. Only the best miners were selected to work here for up to 2 years. Payment was good by Soviet standards. This concept worked perfectly well and it is a good example of how to design a human outpost.

4.9 Robots

Humans have been dreaming of humanoid robots for centuries. The appearance of humanoid robots tries to imitate the human body as much as possible. In science fiction, humanoid robots may be indistinguishable from real humans (in this case, the correct term is android). In real life, humanity is far away from being able to create machines that are indistinguishable from real people and until the first android will be mistaken for a human. In the past century, advancements have been made in the field, in particular with regard to gait and motor skills. Technology now allows for robots to have an independent power source, carry items, get up again when they have fallen, play music instruments and have a conversation. In addition, a lot of progress has been made in the field of artificial intelligence. Learning abilities and autonomy have evolved quickly. It is possible to have a simple conversation with such a system that looks at the conversation partner, follows them with the eyes and shows proper mimical expressions. The boundaries of

Fig. 4.22 Abandoned mining settlement Pyramiden on Svalbard in the Arctic (Images: B. Ganse)

verbal communication and understanding are massive, though. In spaceflight, humanoid robots are used to help with the exploration of the Solar System, repair and service of spacecraft, dangerous spacewalks and to serve as an entertaining team member. NASA has developed the *Robonaut*, a *Robotic Astronaut* for spacewalks. The current version, that has been on board the ISS between 2011 and 2018, is Robonaut 2 (Fig. 4.23). This version is able to use tools with its hands. It is currently only used inside the ISS and not ready for EVAs yet. Several universities are involved with projects on robots in weightlessness and social behaviour issues of humans in contact with them. Their influence on crew psychology seems to be positive. In August 2019, the Russian humanoid robot FEDOR was brought to the International Space Station for a short visit. It demonstrated the ability to manipulate objects while under remote control from the ground.

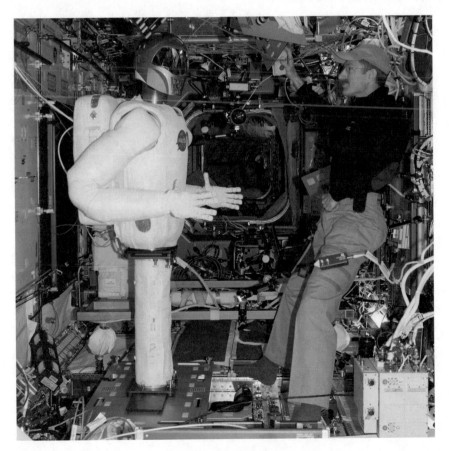

Fig. 4.23 Robonaut 2 and US astronaut Dan Burbank on board the ISS (Image: NASA)

Humanoids are joined by a growing fleet of floating robot companions. The flying mini-satellites called *SPHERES* (Fig. 4.24) have been on the ISS since 2006. They are equipped with CO_2 propulsion, to experiment on autonomous coordination and navigation as a group in microgravity. The newer *Astrobees* expand on their capabilities, including the possibility to hold on to surfaces with a robot arm and move by swinging. Similarly, the robot *CIMON* (Crew Interactive Mobile Companion) couples artificial intelligence with autonomous flying in weightlessness and is used on board the ISS. It was developed in a partnership of Airbus, IBM, LMU Munich University and the German Aerospace Center (DLR). The acronym CIMON refers to the flying brain of the anime series *Captain Future*. Yet another robot currently on board the ISS is the *Int-Ball*, a mainly Japanese project controlled by JAXA (Japan Aerospace Exploration Agency) scientists since 2017. Its primary function is photo and video documentation.

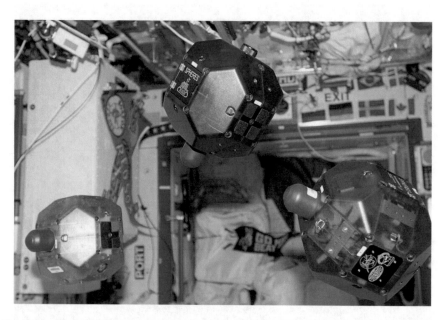

Fig. 4.24 Three SPHERES on board the ISS (Image: NASA)

References

Gushin, V. I., Zaprisa, N. S., Kolinitchenko, T. B., Efimov, V. A., Smirnova, T. M., Vinokhodova, A.G., et al. (1997). Content analysis of the crew communication with external communicants under prolonged isolation. *Aviation, Space, and Environmental Medicine, 68*(12), 1093–1098.

Jemison, M. (2003). *Find where the wind goes.* New York: Scholastic Press. ISBN:978-0-439-13196-0.

Lewin, K., Lippitt, R., & White, R. K. (1939). Patterns of aggressive behavior in experimentally created social climates. *Journal of Social Psychology, 10*, 271–301.

Mora, M., Wink, L., Kögler, I., Mahnert, A., Rettberg, P., Schwendner, P., et al. (2019). Space Station conditions are selective but do not alter microbial characteristics relevant to human health. *Nature Communications, 10*(1), 3990.

NASA (1970). *Apollo operations handbook extravehicular mobility unit* (Vol. 1). System Description Apollo 14. http://www.hq.nasa.gov/alsj/A14EMU-v1.pdf

5

Space Medicine

Contents

Some medical details were already mentioned in previous chapters, but now it is time to get a systematic overview of an exciting field that is essential for the survival of spacefarers. Many body functions and adaptation processes in space are unlike what we know from Earth and often counter-intuitive. When planning a spaceflight, medical aspects should not be neglected. Strong G-forces at start and landing, weightlessness, isolation and exposure to a misanthropic environment lead to a number of changes in the human body. These may cause symptoms and diseases, or sometimes even have consequences for health later in life. This chapter is meant to prepare for these problems and serves to make spaceflight candidates aware of signs and symptoms. It helps to be ready for issues like the elongation of the spine by 5 cm and more, to recognise G-measles, to know what to do in case of space sickness and which symptoms typically show up when the air composition is not right.

© Springer-Verlag GmbH Germany, part of Springer Nature 2020
B. Ganse, U. Ganse, *The Spacefarer's Handbook*, Springer Praxis Books,
https://doi.org/10.1007/978-3-662-61702-1_5

What are humans facing during a spaceflight? The following items are of particular importance:

- Limited resources (food, drinks, etc.)
- Isolation, confinement and a tight time schedule
- Weightlessness
- Gases/poisonous substances/smells in an artificial environment
- Noise
- Hyper-G (acceleration) during start and landing
- Radiation

Figure 5.1 gives an overview of adaptation processes of the human body in weightlessness. Space Medicine is still a young medical discipline and so it makes sense to reflect on its history. For this reason, the chapter starts with how things evolved, followed by a closer look at several organ systems, diseases and symptoms.

Adaptation of the body to weightlessness

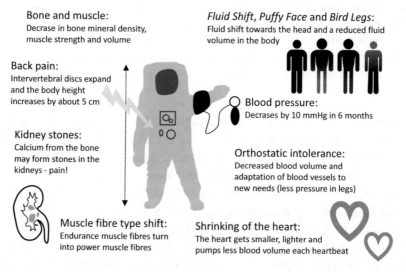

Bone and muscle:
Decrase in bone mineral density, muscle strength and volume

Fluid Shift, Puffy Face and Bird Legs:
Fluid shift towards the head and a reduced fluid volume in the body

Back pain:
Intervertebral discs expand and the body height increases by about 5 cm

Blood pressure:
Decrases by 10 mmHg in 6 months

Kidney stones:
Calcium from the bone may form stones in the kidneys - pain!

Orthostatic intolerance:
Decreased blood volume and adaptation of blood vessels to new needs (less pressure in legs)

Muscle fibre type shift:
Endurance muscle fibres turn into power muscle fibres

Shrinking of the heart:
The heart gets smaller, lighter and pumps less blood volume each heartbeat

Fig. 5.1 What are the main challenges of the human body in weightlessness (not including long-term issues)?

5.1 History of Space Medicine

In the 1950s, when many doctors and scientists due to their research and experience already knew humans could survive a spaceflight, the public was worried and rumours circulated that survival in weightlessness and space radiation was impossible. The strangest theories spread in the news. It was a similar situation to when the first trains and cars appeared in the late nineteenth century when people thought the speed might kill them. In the 1950s, news articles expressed concerns regarding death from acute radiation disease due to huge radiation levels in Earth orbit. The fact that precise measurements of radiation levels were not available yet (Sect. 5.7) opened room for speculation. Newspapers even assumed blood circulation may come to a standstill, resulting in immediate death. This theory did not make any sense at all to anyone more involved with the topic, but that did not stop rumours from spreading. Another story that arose was the gastrointestinal tract coming to a halt, which would make intravenous nutrition necessary. In summary, the field of Space Medicine was surrounded by speculation and controversies up to the year 1957.

In that year, the Soviet Union broke all speculations by sending a dog, *Laika* to the Earth orbit on the Sputnik 2 mission. The female dog was monitored with telemetry to study the effects of spaceflight on its body. In those days, animal experiments were less questioned than today. As the technical capability to safely land back on Earth was not available yet, certain death of the dog was expected and accepted. It was then a great surprise to many that Laika initially survived the launch and some time in weightlessness. For many years the story was upheld that she died from a shortage in oxygen, as an insufficient amount was brought along. This was later corrected by Russian space officials: now overheating had apparently been the true cause of her exitus. Whatever it was, the Sputnik 2 mission has demonstrated the ability of animals to survive a launch and the microgravity environment. Many more animal experiments succeeded, and in 1960, the Soviet Union brought the first dog from orbit back to Earth alive. At the same time, the US were running similar experiments with apes in early project Mercury and came to similar findings.

The space community now felt confident to send a human to space. The initial generation of crewed spacecraft were Vostok in the Soviet Union and Mercury in the USA. Both nations wanted to demonstrate superiority by being the first to bring a human to space—the *Space Race* had begun. This time period was characterised by the pure desire to succeed and to earn political prestige, rather than the gain of scientific or medical knowledge. The race was won by the Soviet Union on 12 April 1961 by sending Yuri Alekseyevich Gagarin to low Earth orbit in Vostok 1 and returning him safely. He was the first human to cross an altitude of 100 km, commonly used as the boundary where space begins. Gagarin orbited Earth once in 108 min. The first American was Alan Shepard on 5 May 1961, who reached an altitude of 187 km before landing in the ocean after 15 min and 22 s. John Glenn was the first American to actually orbit the Earth in 1962. He completed three orbits in the Mercury-Atlas-6-Mission called *Friendship 7*. In 1998, he returned to space again on board the Space Shuttle *Discovery*, becoming the oldest person to participate in a spaceflight so far at age 77 years (as of 2020).

When the human ability to fly to space and survive was no longer a question, the Soviet Union sent up the first woman in Vostok 6 (1963). *Valentina Tereshkova* was the only Soviet woman in space until Svetlana Savitskaya followed her in 1982 and 1984, becoming the first woman to perform a spacewalk. The first American female astronaut Sally Ride went to space in 1983.

The following medical results of the Mercury programme were published (Johnston et al. 1975):

1. Humans can exist in a spaceflight environment.
2. No signs of impaired performance were observed in the astronauts.
3. All measured physiological parameters were in their normal ranges.
4. There were no indications of sensory or psychological issues.
5. The radiation dose was clinically insignificant.
6. After return to Earth, an increased heartbeat and decreased blood pressure were observed.

Six weeks after Yuri Gagarin's flight, John F. Kennedy announced the *Apollo programme* and to put a man on the Moon. To prepare, ten *Gemini-flights* were planned. These were intended to develop and test the technologies needed. To do so, not only navigation and docking manoeuvres, but also overall *operational proficiency* were the aim. For the first time a larger set of medical experiments was run. The Gemini programme involved a total of over 2000 h of human presence in space (Fig. 5.2). Another new feature were extensive *pre-, in- and postflight-experiments* conducted systematically. Blood and urine samples were taken and analysed, opening a whole new chapter of medical research.

Fig. 5.2 The astronauts of Gemini 8 in their spaceship after landing in the Pacific Ocean. On the right-hand side, Neil Armstrong, who later became the first man on the Moon, sits in the capsule. On the left, David Scott is shown who later walked on the Moon with Apollo 15. To be able to find the capsule in the water, fluorescein was used that has a shiny green appearance and improves visibility by search helicopters. The spacecraft could hardly be smaller to host two humans (Image: NASA)

Among the many medical findings of the Gemini programme, the following were considered the most significant:

1. Weight loss and low blood pressure with a tendency to faint were observed after flight. This outcome was later related to decreases in blood volume and weakening of vascular reflexes (more in Sect. 5.2).
2. After 8 days in space, a 20% loss of red blood cells was found. These cells are responsible for the transport of oxygen and carbon dioxide in the body. In addition, altered concentrations of sodium and potassium showed up in the blood. In later spaceflights, these results could not be reproduced anymore, and it is believed that the 20% loss of red blood cells and changes in sodium and potassium concentrations were due to excessive G-exposure of

the astronauts during landing. The impact led to haemolysis, a destruction of the blood cells, that released potassium from the cells' inside to the blood stream. With better parachutes and improved technology, luckily haemolysis does not occur anymore at landing, but in case of a hard landing with high G-forces, it needs to be expected again.

3. The first spacewalk of an American was associated with particularly high calorie consumption.
4. For the first time, a loss in bone mineral density was measured (more on demineralisation of bone in spaceflight in Sect. 5.3).
5. Neither psychological issues nor problems with the sense of balance were found.
6. During Gemini 10, one of the astronauts experienced signs of decompression sickness. Among these symptoms were the bends, joint and muscle pain related to nitrogen bubbles forming in the tissue and collecting in the joints. Nitrogen is usually evenly distributed in the body tissue, but forms bubbles when the pressure suddenly drops, just like when opening a bottle of sparkling water (more on this in Sect. 5.5).

Parallel to the Gemini programme, the Soviet Union drove their development forward in the *Voskhod programme*. On 18 March 1965, Alexei Leonov accomplished the first spacewalk in Voskhod 2. Again, the USSR was ahead of the Americans. Voskhod had a much higher amount of payload compared to its predecessor Vostok, carrying two cosmonauts at a time. Two manned flights were performed. For the first time, the *Space Adaptation Syndrome* was reported: In weightlessness, the brain receives conflicting information from the eyes and the inner ear, leading to nausea and vomiting. Symptoms usually disappear after a few days in weightlessness (Sect. 5.6).

The next goal of the *Space Race* was the Moon! Both nations had a Moon programme and were aiming at getting the first person there. For the *Apollo programme*, the US built the largest rocket in human history: the *Saturn V*. More precisely, as of early 2020, it remains the most powerful, tallest and heaviest rocket in human history. Designed by the German engineer *Wernher von Braun*, the Saturn V was 110.6 m tall. After flights with smaller Saturn 1 and Saturn 1B rockets, the first crewed flight around the moon was carried out in 1967 and the first human landed on the Moon with Apollo 11 in 1969. The Moon landing was a gigantic media spectacle with the greatest possible outreach and enthusiasm. A total of six crewed Apollo Lunar Modules landed on the Moon and twelve Americans walked on its surface. Until today, only these twelve US astronauts have been on the Moon, the last one in 1972 (Fig. 5.3). This was the only time in human history so far, that humans landed

Fig. 5.3 James Irwin on the Moon during the Apollo 15 mission. The lunar lander can be seen on the left (Image: NASA)

on the Moon. The Soviet Space Program never sent humans to the Moon, despite their efforts. The USSR developed the N1 rocket, a counterpart to Saturn V. To make it short: all four uncrewed launch attempts failed, the second causing one of the largest non-nuclear explosions in history, destroying the entire launch complex. Apart from not landing a person on the Moon, the programme did also not reach the stage of bringing cosmonauts outside of the low Earth orbit or around the Moon. No other nation has attempted a crewed flight to the Moon since. In the near future, there will be no living human on Earth, who has walked on the Moon.

The large Apollo Biomedical Research Program led to impressive new findings. A very detailed and informative book was published by NASA

Fig. 5.4 Book *Biomedical Results of Apollo*, NASA SP-368, http://history.nasa.gov/SP-368/contents.htm

featuring all medical details of each mission: *Biomedical Results of Apollo* (Johnston et al. 1975, Fig. 5.4). This is an excellent read for those with an interest in space history.

Apollo's biomedical programme had three goals (in this order):

1. Crew safety and health with a focus on *inflight illness*.
2. Prevention of contamination from possible unknown extraterrestrial organisms.
3. Research on the normal physiological effects of spaceflight on the human body.

The main results included considerable constraints by the Space Adaptation Syndrome, a reduced cardiac stroke volume (the amount of blood the heart pumps in one go), reduced parameters of exercise endurance after the flight, weakening of the heart and blood circulatory system, weight loss of 0.5–6 kg and changes in the concentrations of several hormones, including renin, vasopressin and aldosterone. The 20% loss in red blood cells found in Gemini could not be confirmed and amounted to 2% in Apollo.

An important element of the programme was prevention of back-contamination of the Earth from the Moon with (unknown) extraterrestrial organisms. After each EVA, the spacesuits had to be cleaned with a vacuum cleaner. Upon return to Earth, the astronauts were isolated in special suits called *BIG* (Biological Isolation Garment). They had to wear it until they had reached the *Mobile Quarantine Facility* (MQF), a mobile unit containing bunk beds and living space that would contain the astronauts during transport by sea, air or road to the *Lunar Receiving Laboratory* (LRL). The LRL also called Building 37 of the *Johnson Space Center* in Houston was specifically built for postflight isolation. Everything that came back from the Moon, including the spaceship, astronauts and samples, was isolated in this building. Glovebox vacuum chambers were available for analysis of lunar rocks and dust. The astronauts were kept in quarantine for 3 weeks. Postflight medical research and checks could be completed under very controlled conditions. The Command Module was transported on board a *Super Guppy* plane and sterilised by formaldehyde (gas) over 24 h. When it became obvious that unknown organisms were not a threat, the procedure was discontinued for Apollo 15.

Anecdote

In the year 1971, a medical problem was encountered on the Moon: astronauts Irwin and Scott experienced the so-called *Apollo-15-Syndrome*. Due to a malfunction of the drinking water supply in the astronaut suits, both astronauts dehydrated during their 7 h long EVAs. This was probably paired with high adrenaline levels and a lack of potassium and magnesium. Both astronauts had pain and swelling of their finger tips and later got arrhythmias (heart problems). Irwin shortly lost consciousness after return to the Command Module due to bigeminy, a type of cardiac arrhythmia (Sect. 5.2).

When the USA had successfully landed on the Moon and public attention was decreasing, the Soviet Union gave up its crewed attempts to fly there. Both nations started to focus on building spacestations and on research in weightlessness. While the Cold War was ongoing, there were also military space activities, such as the Soviet military *Almaz* spacestations. The USSR operated their Salyut spacestations from 1971 to 1991, followed by the spacestation Mir (1986–2001). Medical research was published sparsely by the Soviets, while the US made most of their research available to the international community. In May 1973, the last Saturn V rocket was used to deliver the spacestation *Skylab* to space, the only purely American spacestation so far (Fig. 5.5). The spacestation comprised of the upper stage of the rocket, where the fuel tanks were used as living space. At that time, Skylab was the biggest spacestation in history and operational from 1973 to 1979. Three crews stayed on board between May 1973 and February 1974, while the station was unmanned in between and after crew visits. A large medical programme was completed on board, using a lot of new technology.

Fig. 5.5 NASA's *Skylab* spacestation (Image: NASA)

Anecdote

A longer operation time was originally planned for Skylab, but water and oxygen had to be refilled for further use. NASA calculated the station would remain in orbit until roughly 1983 and planned to use a Space Shuttle to bring a propulsion module up that would stabilise the orbit and keep the station going. The Space Shuttle mission STS 2a was scheduled to complete this mission in 1979. In 1978, contact to Skylab was re-established and uncontrolled rotation of the station was discovered. The central computer on board was still working sufficiently, but attitude control had failed. Skylab lost altitude faster than calculated due to higher solar activity making the atmosphere expand and reach higher into space. The air particles led to increased resistance and made Skylab drop. In December 1978, NASA announced they were unable to rescue Skylab, as the Space Shuttle was not operational yet, and no other spaceship was available at that time. On 11 July 1979, Skylab deorbited and broke into many pieces only 16 km above the Earth. It crashed and parts were spread over a large area of land southeast of Perth, Australia. Luckily, no human casualties were reported, but apparently

(continued)

a cow was hit and passed away. The municipal government of Esperance in Australia fined NASA 400 Australian Dollars for littering. NASA, however, never paid the bill. Parts of the spacestation are exhibited in the Esperance Municipal Museum.

Medical experiments on Skylab included blood tests, mineral balance, bone density measurements, sleep- and balance studies, as well as body mass measurements (in weightlessness with an oscillating chair). In addition, cardiovascular research was conducted using a cycle ergometer and an LBNP = *Lower Body Negative Pressure* device on board (Fig. 5.6). The LBNP device was a chamber applying underpressure to the lower body parts of spacefarers to challenge the cardiovascular system (more in Sect. 5.2). Scientific results were published in the books (two volumes) *Biomedical Results from Skylab* (Fig. 5.7, Johnston 1977).

In the following years, medical studies and scientific questions increased in complexity. At the same time, international cooperation was intensified, as it allowed for more interdisciplinary projects on a larger scale. The first cooperation of the Cold War enemies USSR and USA was the *Apollo-Soyuz Test Project* in 1975. An Apollo and a Soyuz spacecraft docked in low Earth orbit using a docking module to provide an airlock between the differing atmospheric conditions of both spaceships. A nitrogen/oxygen atmosphere was and still is used in the Soyuz spacecraft, while the Apollo command module had a much lower pressure with pure oxygen. After the fall of the Iron Curtain, the Russian spacestation Mir was opened for international cooperation. Space

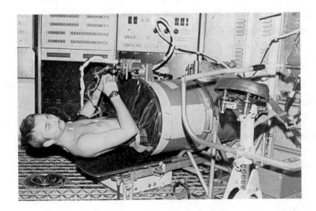

Fig. 5.6 An LBNP (*Lower Body Negative Pressure*) chamber on board spacestation Skylab (Image: NASA)

Fig. 5.7 Book *Biomedical Results of SKYLAB*, NASA, http://lsda.jsc.nasa.gov/books/skylab/biomedical_result_of_skylab.pdf

Shuttles were able to dock to Mir in the *Shuttle-Mir programme*. Spacefarers of several nations worked on joint research projects.

The Space Shuttle offered a unique research environment, called *Spacelab*. As racks and built-in research devices could be reconfigured for each mission, the Spacelab was especially adaptable. Spacelab was built by ten European countries. It was a research laboratory carried in the Space Shuttle cargo bay and used 22 times. A lot of very diverse medical and physiological experiments were conducted in this laboratory, which further improved the understanding of Space Medicine.

Spacestation Mir was followed by the International Space Station ISS, the biggest spacestation so far with impressive research capabilities. As of 2020, the ISS is operational and continuously crewed. In the meantime, China built its first spacestations and currently aims at building a continuously crewed station consisting of several modules. Their first spacestation, *Tiangong 1* was launched in 2011 and deorbited uncontrollably in 2018, while *Tiangong 2* was launched in 2016 and deorbited according to plan in 2019.

From a medical perspective, preparations for long-term missions to Mars or asteroids are currently the focus. Several medical problems need to be addressed before such missions, including radiation protection, medical treatment, neurological and vision problems, muscle loss. These developments were previously given a lower priority, as return from the ISS was always possible within a few hours. The next sections will take a closer look at what exactly happens in the human body during a spaceflight.

5.2   Heart and Circulation

The cardiovascular system (cardio = heart, vascular = blood vessels) is of particular importance to spaceflight, as weightlessness leads to significant changes that put spacefarers at risk. These changes include a redistribution of fluids, altered blood pressures, reduced exercise endurance, arrhythmias and a shrinking heart. It is advisable to be aware of these issues to be able to interpret signs and symptoms correctly.

Fluid Shift, Bird Legs and Puffy Face-Bird Legs Syndrome

On the ground, the pull of Earth's gravity constantly draws the blood and body fluids into the lower body parts. A large portion of the venous blood pools in the leg veins for its transport back to the heart. To pump the blood back up again against the pull of gravity, muscles act as a *muscle pump*. In addition, reflexes of the blood vessel walls allow the vessels to contract and make sure blood pooling due to dilation is reduced. The muscle pump is not an organ by itself, but leg muscle contractions push the blood upwards, for example, when walking. Venous valves make sure the blood does not flow straight back down again. Under the influence of gravity, on the one hand, the blood from the head passively flows back to the heart. In weightlessness, on the other hand, spacefarers usually have a swollen face and very thin legs. Here, due to the lack of gravity, the blood effortlessly flows to the heart and headward by itself. Similar conditions show up during a handstand on Earth. The swollen face is called a *Puffy Face*.

Apart from the blood flowing back from the legs to the heart easily, another phenomenon contributes to the thin legs: the lack of *hydrostatic pressure*. Gravity usually causes hydrostatic pressure to build up in the column of blood in a blood vessel, leading to fluid (but no blood cells) leaking out of the small blood vessels where the wall is thinnest and the pressure highest. This fluid in the tissue usually makes the legs appear a bit thicker, depending on individual predisposition. In space, this additional fluid will never leave the blood vessels and the legs stay thin. The official terminology for thin legs in space is *Bird Legs*. Together, the Puffy Face and Bird Legs form the *Puffy Face-Bird Legs Syndrome*.

When fluid collects in the head, the blood pressure sensors (called baroreceptors) in the neck and Aorta blood vessel walls sense higher pressures and activate the *baroreceptor reflex*, or baroreflex. A nerve signal is sent to the brain, letting it know the blood pressure is too high. In response, the *renin–angiotensin–aldosterone system* (RAAS) is activated by the brain. By regulating hormone concentrations, the kidney is instructed to release water. Due to this mechanism, 1.5 l of water are usually excreted in the first 24 h of a spaceflight to regulate the fluid shift. This leads to a lower total blood volume compared to before launch, while the number of blood cells remains the same. The volume percentage of blood cells in the blood, the *haematocrit*, increases. Consequently, it slowly normalises again over the course of several weeks in space. In the later stages of a spaceflight, the *Puffy Face* is less pronounced than before and the amount of urine excreted returns to normal. On return to Earth, when gravity again pulls blood in the direction of the legs, the baroreceptors sense a too low blood pressure. The RAAS will now work in the other direction. Spacefarers returning from a spaceflight are thirsty as their body longs for water to refill the missing blood volume. Accordingly, the haematocrit drops and new blood cells need to be produced.

A stay in space also leads to habituation of the baroreceptors to higher pressures. After return to the Earth, larger pressure changes are necessary to activate the baroreflex compared to the time before the flight. This practically means that blood pressure regulation is a bit rough and lacks fine-tuning during the first days after landing. It is not uncommon for spacefarers to faint when trying to stand for a longer period of time during this phase. When the blood collects in the legs after landing, it is hard for the circulatory system to keep up with this challenge. The fluid shift is illustrated in Fig. 5.8. The mechanism of fluid regulation is depicted in Fig. 5.9.

Why People Need to be Carried After Landing

The more time one spends in weightlessness, the greater the adaptation of the cardiovascular system to a less demanding lifestyle. The technical term is cardiovascular deconditioning. Coming back from a spaceflight, blood collects in the lower extremities and the brain may not get a sufficient oxygen supply. The spacefarer might experience an orthostatic dysregulation with dizziness, heart palpitations, sweating, impaired vision and/or ringing ears. In case of insufficient blood supply to the brain, the field of vision narrows, one might experience a *greyout* (loss of colour vision), and if it gets worse a *blackout* (total loss of vision). Loss of consciousness often follows a blackout and is associated with a risk of injury due to a so-called circulatory collapse (also called *syncope*). In addition to these cardiovascular changes, neurosensory coordination problems lead to trouble balancing, nausea and difficulties to

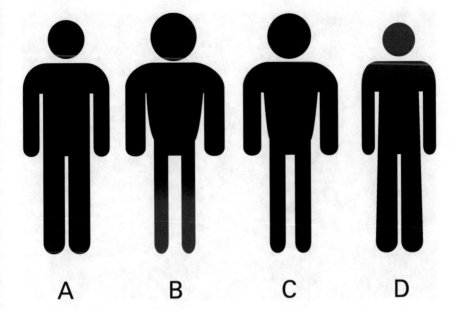

A   B   C   D

Fig. 5.8 Illustration of the fluid shift with *Puffy Face-Bird Legs Syndrome* and *Bird Legs*.
(**a**) before flight, (**b**) initial phase in weightlessness, (**c**) adapted to weightlessness and
(**d**) immediately after landing

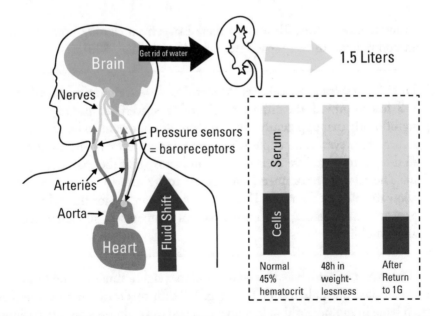

Fig. 5.9 In weightlessness, the *fluid shift* leads to an increased pressure at the pressure
sensors (baroreceptors). The information is forwarded to the midbrain via nerves. The
brain secretes hormones to make the kidneys release more water. The box shows
changes in blood composition as they would show up in blood vials

Fig. 5.10 Typical picture of sitting spacefarers after their return in a Soyuz spacecraft in the steppe of Kazakhstan. Shown here is ISS expedition 20 in October 2009 (Image: NASA)

coordinate movements. To prevent these issues from causing problems, it is common practice to carry spacefarers after landing (Fig. 5.10).

Hint for Aspiring Spacefarers Modern exercise training on board the ISS has improved the fitness of returning spaceflight participants so significantly that apparently some of them did not want to be carried on return. However, the people who were selected to carry the crew were so disappointed to lose their moment in history that the spacefarers gave in. The authors recommend the readers to be nice to the carriers and allow them to do their job. This is not only rewarding for them, but also prevents embarrassing and maybe painful moments of sudden syncope and stumbling.

The extent of cardiovascular adaptation to weightlessness can be measured in a so-called tilt table test (TTT) (Fig. 5.11). In this test, a person who has been lying in supine position is suddenly raised to an almost standing position

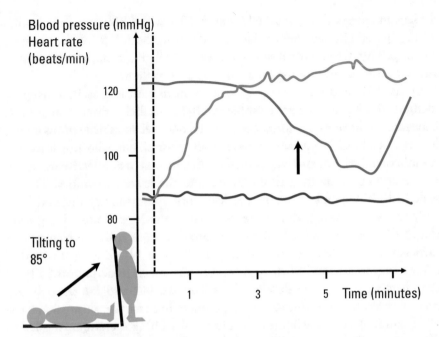

Fig. 5.11 Tilt table test (TTT). The participant is tilted from supine to an 85° half-standing position. The blood pressure (systolic in light blue, diastolic in dark blue) and the heart rate (red) are shown demonstrating the case of a presyncope. The arrow marks the moment when the test should be terminated to prevent syncope. In case of syncope, the heart rate drops

of 85°. A tilt of 85° means that the person still leans against the stretcher and is not tempted to use the muscle pump. The time until orthostatic problems show up after tilting reflects the level of adjustment. Blood pressure and heart rate are monitored. The experiment is interrupted at onset of presyncope, which means the blood pressure drops and heart rate increases. Immediately, the stretcher is tilted back. This imminent orthostatic dysregulation is called presyncope. A higher level of adjustment to weightlessness leads to a shorter time span until presyncope. This experiment is an objective measure of how well someone has exercised in space, and also reveals if a participant has cheated during a bedrest study. Everyone gets a presyncope when only standing long enough without using the muscle pump. In well-trained individuals who have not recently returned from a spaceflight, it would, however, take several hours until problems show up. To make sure the TTT does not take that long, it is common to apply defined levels of underpressure via LBNP to the lower extremity of participants following a standardised protocol. Application of the LBNP device further increases the collection of blood in the legs and presyncope is reached faster. A TTT with LBNP is used in studies testing spaceflight analogues (such as bedrest, see Sect. 5.9) to demonstrate comparability of the study results with real spaceflight. It may also be used to assess the efficacy of training interventions in spaceflight.

In the 1970s and 1980s, space agencies went to great lengths in developing methods to reduce orthostatic problems after spaceflight. Even though these issues are no threat to mission success and no one ever sustained serious injury, it was considered important to have healthy astronauts who can leave the capsule walking by themselves. A method that proved to be effective was again to use underpressure to suck the blood into the lower extremities. The air sealed LBNP chamber contains the legs until approximately the mid-belly and, roughly speaking, draws the blood into the legs, almost like gravity (Fig. 5.6). The cardiovascular system is forced to work against it to maintain sufficient blood supply to the brain. The effect of LBNP is so strong that it may lead to syncope when applied with too much underpressure. LBNP chambers were not only developed for exercise, but also for research. An example was the microneurography experiment in Space Shuttle mission STS-90 (Neurolab). As many hours of cardiovascular LBNP exercise are needed to decrease postflight orthostatic problems, while the device is relatively heavy and big, it is not used as a regular *countermeasure* anymore. Reasons include the fact that exercise methods have improved significantly, also reducing the orthostatic problems. In addition, postflight orthostasis has been given a lower priority. Russia currently has an LBNP on board the ISS for research. It is called the Chibis suit and is designed as rubber pants.

Sex differences exist with regard to orthostatic problems. When blood pressure drops, men react more by contraction of the smallest peripheral blood vessels (when the volume decreases, pressure increases), while women rather react with an increase in heart rate (more blood pumped = higher blood pressure). Both mechanisms lead to an increase in blood pressure.

Reduced Endurance and Shrinking Hearts

After return from space, orthostatic dysregulation is not the only observed change of the cardiovascular system. Spending time in weightlessness without exercise will also lead to shrinking of the heart and to reduced endurance. The term *endurance* means the ability for physical work over a longer period of time. Cells adapt to endurance exercise by increasing numbers of enzymes and structures needed to supply the muscles with energy, for example, creating more mitochondria. Mitochondria are the power house of the cells, providing energy for contraction. On the other hand, during immobilisation, the body gets rid of these structures, as it only keeps what it needs.

Endurance is commonly measured by physiologists, physicians and sport scientists by assessing the maximal oxygen uptake during a run on a treadmill or while cycling on a cycle ergometer. Depending on the number of mitochondria, the amount of oxygen that can be metabolised by the cells varies. The scientific term for the maximum oxygen uptake is VO_2 max (V for volume and O_2 for oxygen). The procedure of measuring VO_2 max is called spiroergometry (Fig. 5.12). Oxygen uptake and carbon dioxide production are measured in the in- and exhaled air. In addition, an ECG (electrocardiogram) and the blood pressure are recorded to guarantee the participant's safety and to stop the experiment when problems occur. The experiment starts with a low load level (wattage). The load is increased according to a defined protocol, so the exercise becomes more and more strenuous. At some stage, the oxygen

Fig. 5.12 Astronaut Sunita Williams performing a spiroergometry on the cycle ergometer CEVIS on board the ISS (Image: NASA)

uptake reaches a plateau and does not increase further. The body has reached its maximum oxygen consumption, meaning it consumes as much oxygen as possible, because the cells are running at their maximal effort (VO_2 max). The second interesting parameter is the load level at which the participant was unable to continue, the maximum wattage, or maximum load. After spaceflight without exercise, VO_2 and maximum load are reduced.

In addition to orthostatic problems and decreased endurance, cardiovascular adaptation to prolonged weightlessness also includes shrinking of the heart, called *cardiac atrophy*. Cardiac shrinkage occurs because the heart pumps against less resistance in weightlessness compared to on Earth. On a planet or moon with more gravity than on Earth, the heart would grow while having to pump harder. In weightlessness the mass of the heart decreases due to thinning of the heart muscle walls. After return to the Earth, the heart re-adapts and slowly increases in mass and volume again. These changes to the heart contribute to the blood pressure and orthostatic problems after return. Over the years, exercise regimes for spaceflight have been improved substantially and help reduce cardiac atrophy. However, as soon as a spaceflight is planned without the option to exercise or the training is not carried out for other reasons, the described problems will present not only when landing on Earth, but also when landing on Mars or elsewhere.

Central Venous Pressure

For a long time, blood pressure changes in weightlessness were a matter of speculation, as precise measurements were not possible. These can only be obtained invasively, which means by having a probe inserted into a vein or artery. German Physiology professor Karl Kirsch in 1984 predicted that venous blood pressure would drop to almost zero in weightlessness and published his thoughts and arguments in the scientific journal *Science* (Kirsch et al. 1984). No one had measured venous pressure in spaceflight until then. The scientific community questioned Kirsch's paper and then mainly ignored it, until the American astronaut, physician and scientist Jay Buckey and colleagues got permission to measure the pressure invasively during a Space Shuttle flight in 1991. In the Spacelab Life Sciences 1 mission, a central venous catheter was inserted into one crew member's arm and moved forward to his heart to measure the *central venous pressure* (CVP). CVP was recorded throughout the entire launch phase until the low Earth orbit was reached. Buckey and his coauthors speculated the pressure would increase, as it does in head-down tilt bedrest on Earth, leading to a *Puffy Face*. During the launch phase, high blood pressure values were recorded while enormous forces and vibrations were acting on the bodies. When the Shuttle entered the orbit, central venous pressure suddenly dropped to almost zero, just as predicted by Kirsch. This was a scientific sensation that got published in the *New England Journal of Medicine* in 1993 (Buckey et al. 1993). The measurements were repeated on two more astronauts during the Spacelab Life Sciences 2 mission, and results again matched the previous finding (Buckey et al. 1996).

The explanation for the pressure being almost zero seems to be that weightlessness reduces resistance and pressure of the surrounding organs and tissues. The organs float. Despite the measured pressure being low, sufficient blood volume flows to the heart. Most pressures in the body change in weightlessness and central venous pressure cannot be used as a parameter to evaluate health, as on Earth. It is currently unknown how these pressure changes will affect the health of humans flying to space with cardiovascular diseases, for example, after a heart attack. So far (as of 2020), nobody with heart problems has been to space. It does, however, seem likely that weightlessness reduces the work the heart has to do, and is therefore most likely bearable for humans with heart disease. Launch and return to gravity, however, would likely pose a major threat to this group.

Arterial Blood Pressure

For individuals with hypertension (increased arterial blood pressure), weightlessness is great! The systolic blood pressure decreases by 5 mmHg in 2 weeks and by 10 mmHg in 6 months. This is a medically significant amount! The decrease in blood pressure seems to be caused by dilation of blood vessels. In weightlessness, small arteries are more expanded than in gravity. A possible mechanism is that activation of the renin-angiotensin-aldosterone system decreases circulating levels of vasopressin.

G-measles

G-measles (G = gravitation) are neither a childhood disease nor infectious or preventable by vaccination. Instead, they form in high G-forces as little blood vessels rupture and release blood that collects in the skin (Fig. 5.13). These blood collections are also called petechiae. G-measles also appear in fighter pilots, car accidents or may even show up in very sensitive individuals after a roller coaster ride. G-measles always appear in the location where the highest

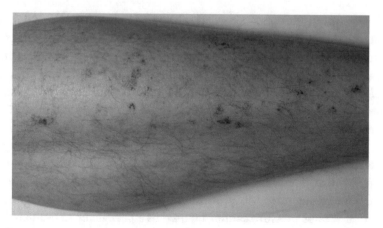

Fig. 5.13 G-measles of the lower leg (Image: B. Ganse)

hydrostatic pressure weighs on the blood vessel walls. Some individuals are more likely to get them than others, and the likelihood and severity increase the longer a person has been weightless. During launch and landing of spaceships like Soyuz or Apollo, the crew lies on their back and G-measles therefore appear(ed) in the skin of the back. These usually show up when landing was tougher than planned. Spaceflight participants should know what G-measles look like to be able to distinguish them from other pathologies. The good news: G-measles usually disappear by themselves and do not require treatment. In severe cases, they may flow together and form larger spots. While fewer G-measles are harmless, it needs to be considered that bigger blood collections are usually not only located in the skin, but may also affect inner organs. Therefore, in case of bigger blood spots in the skin, further diagnostic steps such as an ultrasound examination of the inner organs are recommended. When the body resorbs the blood, it may change colour to green and yellow.

Irregular Heartbeats

Irregular heartbeats (arrhythmias) were a more frequent problem in the early years of spaceflight compared to today. During the Apollo missions, arrhythmias occurred in the orbit around and on the Moon, as well as on return to Earth. Arrhythmias were also reported during the Salyut 7 and Skylab missions. On the spacestation Mir and on some Space Shuttle flights, arrhythmias happened during EVAs. The events are usually triggered by very strenuous work, exposure to high G-forces or dehydration. In case of extreme stress, the same is observed on Earth: the heart muscle temporarily malfunctions. This phenomenon is called *stress cardiomyopathy*. On the spacestation Mir, in times of insufficient exercise, it was also reported that more pronounced cardiac atrophy increases the likelihood of arrhythmias. With higher G-forces, for example, in human centrifugation, arrhythmias are seen more frequently and they usually disappear when the G-exposure stops (more on human centrifugation in Sect. 5.9). Furthermore, additional heartbeats are frequently observed, so-called extrasystoles. The most common type of arrhythmia in spaceflight is called *ventricular tachycardia* (VT, Fig. 5.14) and characterised by a fast heart rate from improper electrical activity in the heart muscles. A VT is potentially life-threatening and needs to be taken very seriously! If the heart is still able to pump a sufficient amount of blood, symptoms are weakness, palpitations, light-headedness and maybe chest pain. Unconsciousness may occur in more severe cases. If the arrhythmia persists after stopping the centrifuge or after landing, it needs medical treatment. It may be necessary to apply an electrical impulse. Luckily, such a case has not yet been reported in human spaceflight history. Another arrhythmia that occasionally occurs in spaceflight is the *bigeminy*. In this case, each normal heartbeat is followed by another heartbeat that comes too early and is therefore ineffective, followed by a pause. Also the bigeminy usually ends when the cause is gone.

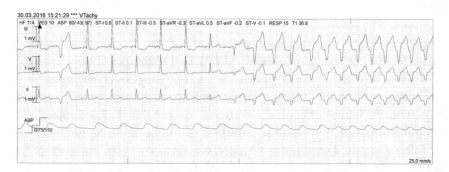

Fig. 5.14 A human ECG (in this case not from a spacefarer) showing typical changes that may also occur in spaceflight. The upper three curves show the ECG of the heart (the electrical activity). Below, the arterial pressure curve is shown. From left to right, the first heartbeat is relatively normal, followed by an extrasystole (an additional heartbeat). After five additional normal heartbeats, the rhythm changes. A ventricular tachycardia may look like what is shown here (Image: B. Ganse)

5.3   The Musculoskeletal System

Bones, muscles, cartilage, intervertebral discs, tendons and fascia form the musculoskeletal system. In weightlessness, these structures do not carry as much load, as no force is needed to move around and carry objects. Everything is weightless in weightlessness! The heaviest item may be moved with ease. To save resources, the body makes sure the unused structures of the musculoskeletal system are reduced in volume and mass. Muscle volume and force decrease rapidly and minerals (calcium and phosphate) are removed from the bones. In addition, the missing load leads to swelling of the intervertebral discs. The spine therefore elongates and surrounding structures are stretched, causing discomfort and back pain. The body posture changes and the spine is straightened more than usual. All these changes are so severe that persons who have spent a longer time in weightlessness without appropriate exercise and bone medication experience problems after landing back on Earth. The loss in bone mass (called osteopenia) slightly increases the likelihood for bone fractures. The loss of muscle mass not only affects strength but also the metabolism. The following pages will explain and differentiate these issues to illuminate an often underestimated topic.

Bone

Bones not only have mechanical properties and functions but also act as an organ for the production of new blood cells and hormones. The blood supply of bones is very high. Several types of bones exist: long bones (for example, the upper arm and forearm bones), short bones (such as the wrist bones), flat bones (e.g. skull and pelvis), irregular bones (example: vertebrae) and sesamoid bones (to guide tendons, for example, attached to the big toe). Some bones may be filled with air. There are two general classes of bone structure: cancellous (spongy, Fig. 5.15) and cortical (very solid). In most cases bone marrow is located inside the long bones and surrounded by cortical bone.

Bone tissue consists of organic (30%) and inorganic (45%) compounds. The organic components include collagen and proteoglycans, while the inorganic portion involves hydroxyapatite, calcium and phosphate. 25% of bone is water. Bone cells have the ability to sense deformation and steer bone tissue formation and absorption accordingly. Osteoblasts are cells that create bone tissue, while osteoclasts remove it. Both cell types are active in parallel, while their activity can be up and down regulated by a complex system of hormones and substances, resulting in a net increase or decrease in bone mass. When a

Fig. 5.15 Cancellous bone of an elephant femur that has been exposed to the environment for a long time (Image: B. Ganse)

certain level of strain acts on bone, it will be built up, while it is broken down if not loaded adequately. In a range in between, nothing happens to the net bone turnover, while osteoblasts and osteoclasts continue to renew the structures by building and removing bone. The maximum bone density is usually reached at around 30 years of age, and it will then gradually decrease throughout the life span. Due to hormone changes, women experience a faster decline during menopause. Calcium and phosphate are disposed of via the kidneys. In case of increased bone resorption, for example, when in weightlessness without proper exercise, kidney stones may form from the excreted calcium. These are not only very painful, they may even lead to urine congestion and infections of the kidneys and urinary tract. If such an infection is not treated properly, be it because antibiotics are not available, it can be very dangerous and maybe even deadly.

Bone mineral density can be measured using osteodensitometry applying X-ray based methods. The two most common techniques are DEXA (Dual-energy X-ray absorptiometry), using two X-ray beams with different energy levels, and QCT (Quantitative computed tomography), computer-processed combinations of several X-ray images taken from different angles. These measurements use the fact that the more X-rays are absorbed by the tissue, the higher the bone density. A T-score is calculated that describes the deviation of the value from the norm. Bone density is reduced when the T-score is -1 to -2.5. In case of values under -2.5, the patient probably has osteoporosis and needs further diagnostic (blood) tests and most likely treatment. Bone density was systematically investigated in NASA astronauts beginning in 1998. Astronauts currently (2020) require a T-value of -1.0 or more to fly. In spaceflight, the monthly loss in bone mineral density is 1–2%, while only a 0.5–1% loss per year occurs on Earth. During a 6 months stay on board a spacestation without proper countermeasures, bone density losses of 15–20% were found in the past. When considering a 2-year mission to Mars, a bone loss of up to 50% might be possible, increasing the risk for bone fractures significantly. Luckily countermeasures are in place to mitigate the effects of weightlessness. The extent of bone density losses varies between bone locations. Those bones lose the most where unloading leads to the greatest load difference compared to before. This is particularly the case in the legs, pelvis and lower spine. Bone density usually does not decrease in the arms, and it may even slightly increase in skull bones. During long stays in space without adequate countermeasures, not only bone density, but also bone geometry changes, such as the pattern of the bone structure.

To stop bone loss, a combination of several countermeasures proved to be efficient. Resistive exercise (Sibonga et al. 2019) plays a central role here, which will be explained in detail in Sect. 5.4. In addition, it has been shown that oral vitamin D3 and calcium in combination with bisphosphonates, a medication class that decreases the activity of bone-resorbing osteoclasts, have helped to solve the bone loss problem in spaceflight (Fig. 5.16). Vitamin D3 plays an important role in bone formation and is usually produced when the sun shines on the skin. As sunbathing leads to sunburns in spaceflight, daily vitamin D3-intake is recommended for spacefarers. In addition, calcium should be taken to ensure sufficient amounts are available to be integrated in the bone structure. Depending on the type of mission, artificial gravity might be available in the future to be applied as additional countermeasure (Sect. 2.2). A combination of the named interventions has recently led to a more or less stable bone density in spaceflight participants who were on board the ISS for 6 months and more. The bone-problem can now be considered roughly solved for spaceflight and it can be assumed that an increased risk of bone fracture will not be a major issue if these countermeasures are properly applied.

Hint for Aspiring Spacefarers To decrease bone loss in weightlessness, a combination of daily resistive exercise with oral vitamin D3, calcium and bisphosphonates can be applied. The combination of these countermeasures has proven to help against the bone loss in spaceflight.

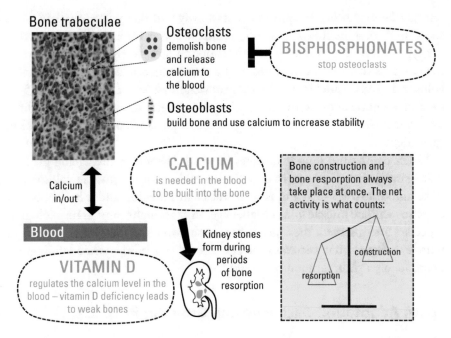

Fig. 5.16 Illustration of bone composition and degradation plus the effects of vitamin D3, calcium and bisphosphonates

Muscle

The skeletal muscles enable animals and humans to move. Consisting of 80% muscle fibres that may be up to 15 cm in length, there are roughly speaking two types of muscles fibres with several sub-groups:

1. Type I: Slow fibres, high endurance, low power, aerobic, many mitochondria, a high concentration of myoglobin
2. Type II: Fast fibres, low endurance, high power, anaerobic, lots of glycogen and enzymes that break down sugar.

Typically, endurance athletes have more type I fibres and power athletes more type II fibres. The distribution of fibres also depends on the muscle and varies depending on the muscles function. A muscle with the function of maintaining posture while standing, such as the deep muscle of the lower leg, has many type I fibres. The power-generating muscles in the upper leg that are needed to sprint usually have more type II fibres. In weightlessness, type I fibres slowly transition into type II fibres. This process is called the *muscle fibre type shift*. Transition occurs slowly and many intermediate fibres show up during the first weeks and months in space. Intermediate means that the type I fibre loses some of its mitochondria and slowly produces more of the enzymes that break down and help to burn sugar. In addition, the muscle volume decreases quickly during a stay in weightlessness. During the first 5–11 days in weightlessness without exercise, the skeletal muscle volume declines by about 20%. This number highlights the importance of muscle exercise in spaceflight.

When comparing muscle changes in weightlessness to the ageing process, many similarities show up. In older individuals, losses in muscle volume and force amount to 1.5–3% annually. Insufficient nutrition and a lack of power exercise facilitate muscle loss at higher age. In spaceflight, similar physiological processes are observed, they just occur much faster and may lead to a level of muscle weakness that decreases the operational capabilities when landing, for example, on a planet, moon or asteroid.

Space Adaptation Back Pain and Growing in Space

In weightlessness, the intervertebral discs (IVDs) expand and become thicker. During the first 24 h in weightlessness, spacefarers grow by 5 cm on average. The reason is that gravity causes the body to compress the IVDs while on Earth. If this pressure is gone, water collects in the IVDs and they become thicker. In addition, the body posture and straightening of the spine differ in spaceflight, adding to the impression of growth. Figure 5.17 illustrates why IVDs swell. Following spinal elongation, approximately half (52%) of the spacefarers experience back pain, called *Space Adaptation Back Pain*. The pain is usually described as mild, but 15% of individuals are so impaired and in pain that they cannot complete the assigned tasks. Another aspect is that sleep problems due to back pain may lead to subsequent limitations the next day. Pain medication is usually used to control the problem. In some cases, stronger pain medication is necessary to complete the mission tasks. After return to Earth, the risk for an intervertebral disc prolapse (herniation of the disc in the spinal canal) is increased. Disc herniations can be associated with numbness and decreased force in an arm or leg. Several countermeasures are available: apart from medication, a specific posture may help reduce the pain, the foetal tuck position. The knees are bent towards the chest to compress the IVDs and achieve relief. In addition, it has been found to be very comfortable and relieving to strap a rubber band around the knees and back at night to maintain the foetal tuck position. The same can be reached by using a sleeping bag to

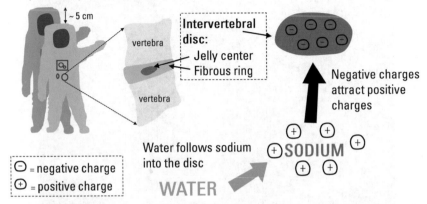

At 1 G, the 23 intervertebral discs are compressed by the body weight

~5 cm

vertebra

Intervertebral disc:
- Jelly center
- Fibrous ring

vertebra

Negative charges attract positive charges

⊖ = negative charge
⊕ = positive charge

Water follows sodium into the disc

WATER

⊕SODIUM

Fig. 5.17 Spacefarers grow in space because their spine loses the S-shape and straightens out. In addition, the intervertebral discs attract sodium and water, leading to an increase in height

keep that position during the night. Another alternative are suits that compress the spine, such as the *Skin Suit* or the *Penguin Suit*. In these, rubber bands help to push the body back together in the longitudinal axis. Human centrifugation might be a good countermeasure against back pain during spaceflight. More on human centrifugation in Sect. 5.4.

Interview with Story Musgrave

Story Musgrave, born 1935, is the only astronaut who travelled on board all five Space Shuttles (Fig. 5.18). He was in space six times between 1983 and 1996. In 1993, he repaired the *Hubble Space Telescope* in a spectacular EVA. Before becoming an astronaut, Story studied Medicine and worked as a surgeon. The authors had the opportunity to talk to him on the phone and record the following interview on the musculoskeletal system in spaceflight.

MUSGRAVE: So, I looked over your form, and if I answered it, you would not get the answers you are after. I can tell you that you are massively vulnerable with back stuff when you come home. I have been on the spot more than once when I saw it happen. You know we grow an inch or an inch and a half in space. You know that already. Whatever the numbers are. Of course, very early on in the Space Shuttle programme, we got into, you know, how much longer you need to make your suit

Fig. 5.18 Surgeon and astronaut Story Musgrave repaired the Hubble-Telescope in 1993. The *Remote Manipulator System* (RMS) carried him to the telescope (Image: NASA)

to accommodate for that. I did not! I let the suit squeeze me down. I let the suit squeeze the water out of my discs.

GANSE: Does that work? Can a tight suit make you shorter up there?

MUSGRAVE: It does! There is no question! From your tip of your toe, every piece of cartilage has no pressure on it, and so it is growing.

GANSE: Does the suit also stop back pain?

MUSGRAVE: I do not know, I never had any back pain. I have not heard of difficulties in the suit. I have not heard of back problems there. The other thing is that if you do not pay attention to it, you are in a very bad posture when you sleep. You sleep in a sleeping bag in a very bad posture. What you need to do is to sleep cross-wise in the bag. You need to get your knees on your chest.

GANSE: I heard some strap a rubber band around their knees and back. Does that work?

MUSGRAVE: Sure, I did that, and I slept floating, too. I did not sleep in the bag. It is very important to stay stretched-out and stay flexible there, so that is an important thing. But I have seen a huge number of cases. I have witnessed cases when people got on the ground that did not look after themselves and who did not understand how fragile they were. I saw them hurt themselves just getting in whatever van, getting in whatever transportation mode you are in, you have to be incredibly careful. I landed out at Edwards once. It was an unexpected landing. We had a problem with the vehicle, so they sent us to Edwards, but they were not prepared to do things. So, we had a very rough van ride to the motel where we were staying. And we did not take the highway. Somebody had a shortcut, and they took us through the dirt road. That was incredibly bad to do. They did not understand how fragile we are. And I have seen people get hurt because they do not understand how fragile they are. They got hurt just bending into the trunk. They did not have to pick up the suitcase. You are unbelievably fragile. I know how fragile my back is at six in the morning as compared to six at night. There is no question about that. I am a landscaper, and architect, I do very heavy work. The heavier work you do, the safer your back is. If you sit around all day, your back becomes vulnerable. But I know, and this is a personal anecdote, looking after my back, I do know that early in the morning when I get up, my back is fragile. Yes, and it seems to be the same fragility, which I see and feel when I come home from a spaceflight. And so, of course, I do not have the exact measurements in my head, but you are half an inch or an inch or so bigger in the morning. They need to do those kinds of studies. …I think that's anecdotally what I got for you.

GANSE: Thank you so much! One more question: how long did it take until your back was normal again after landing?

MUSGRAVE: I am pretty much adapted before we land, just the one G of reentry. But I did a standing-up reentry in the Shuttle. That acceleration was a lot of stress and strain to get photographs of the plasma coming home. But I am sensitive, you see, because I have seen so many other people get hurt. So I do not get myself a chance to get hurt. I know when I feel it. Early in the morning, you gotta look after yourself. You need to get it warmed up and shortened up.

GANSE: Did any of your colleagues get disc herniations after spaceflight, and do you think their behaviour was different?

MUSGRAVE: I do not know of a single disc herniation. But I am anecdotal.
It is just the people I flew with. You have the data—I do not have the
data. I do not collect the data, I am the data!

Anecdote

After return from a spaceflight, the body feels quite vulnerable and one should
be very cautious not to get injured. Astronaut Story Musgrave once experienced
reentry in a Space Shuttle in a standing position. For safety-reasons, this was
strictly forbidden and not approved by NASA. However, he got up anyway to
take photos of the plasma that surrounds the spaceship and only admitted
his disobedience many years later. He was lucky and did not get hurt, which
documents the amazingly smooth landing characteristics of the Space Shuttle.

5.4   Physical Exercise in Spaceflight

As the previous sections showed, exercise countermeasures play a very important role during space missions to protect cardiovascular and musculoskeletal
health. The negative consequences of not exercising in spaceflight have already
been reported. Now it is time to discuss how to exercise to achieve the best
results in the most efficient way. Exercise aims at the following goals:

1. To keep muscle mass and muscle force,
2. To keep bone mass,
3. Prevention of cardiovascular deconditioning, which means in detail

   (a) To keep endurance (VO_2 max),
   (b) To prevent the heart from shrinking,
   (c) To prevent *orthostatic intolerance*.

Developing exercise methods for weightlessness was not at all trivial and
easy. As objects weigh nothing, it is possible to balance the heaviest object on
a single finger. For this reason, classical resistive exercise with weights is not
an option in spaceflight. Also running on a treadmill without being pulled
back by rubber bands is impossible without gravity, as one immediately floats
away to the opposite wall. Scientists for the first time thought about exercise in
spaceflight while planning the Apollo programme. In Mercury and Gemini,
cardiovascular and musculoskeletal deconditioning had been observed, and
regarding the length of the Apollo missions, countermeasures were considered

an important idea. The first exercise device in spaceflight history was the *Exer-Genie*, an off-the-shelf variable resistance rope friction device that is very lightweight and small. Apollo astronauts were allowed to use it as much as they liked without standardisation, which made a scientific analysis of the effects impossible. The crew apparently reported positive feedback for the exercise, so systematic countermeasure research was initiated for the Skylab missions. The first Skylab crew used a bicycle ergometer, the second added a resistive power device, and on top of these two devices, the third crew had the Teflon-treadmill mentioned before (Fig. 5.19). In each mission, leg mass and power improved, while the treadmill showed the strongest positive effect. Due to the great results, NASA decided to build a proper treadmill with a number of functions. The first such treadmill with electronic control and resistance was tested in Space Shuttle flights during the 1980s. Exercise methods were further improved on the spacestation Mir and later on the ISS, where significant progress was made. Currently (in the year 2020), each crew member on board the ISS is scheduled for 2.5 h of physical exercise daily. 1.5 h comprise the actual exercise, while 1 h is needed to get ready and for hygiene afterwards. A combination of power and endurance exercise has proven to be ideal as a countermeasure in spaceflight.

For power exercise, a device called ARED (*Advanced Resistive Exercise Device*) is used on board the ISS that had many generations of predecessors

Fig. 5.19 The treadmill on the spacestation *Skylab* consisted of a Teflon plate and rubber bands. The astronauts had to wear slippery socks to use it, while the rubber bands would pull them towards the Teflon plate. This construction was a milestone is space exercise research (Image: NASA)

(Fig. 5.20). ARED works with vacuum cylinders that provide constant resistance. In addition, a flywheel provides inertial forces to make the experience in weightlessness even more realistic. Many exercises are possible with the device, including dead lifts, squats and calf raises. Not only the muscles are trained by this device, but forces also deform the bones. Bone deformation is the stimulus for the bone to increase bone density to improve stability. As mentioned before, the ARED combined with specific medications has stopped bone loss in weightlessness—a great scientific achievement!

For endurance exercise, the ISS provides treadmills (Fig. 5.21) and a bicycle (Fig. 5.12). Each crew member uses the treadmill 4–6 times and the bicycle

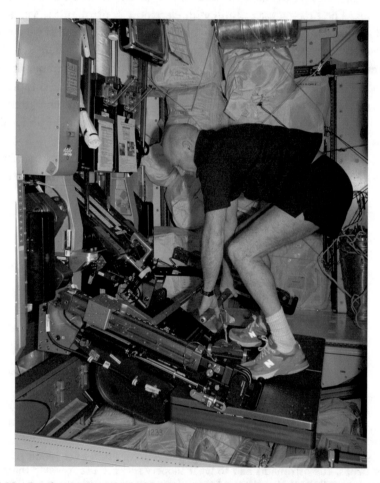

Fig. 5.20 Dutch astronaut André Kuipers using the *ARED* (Advanced Resistive Exercise Device) on board the ISS (Image: NASA)

Fig. 5.21   Sunita Williams on the treadmill (COLBERT) on board the ISS (Image: NASA)

2–3 times per week. Each session is 30–45 min long at 70–75% of their maximal heart rate. The bike is called CEVIS (*Cycle Ergometer with Vibration Isolation and Stabilization System*). Vibration isolation means that movements and vibrations of the device are not transferred to the spacestation structure. This is to make sure experiments can run smoothly and forces do not put strain on the station structure. Exercise devices therefore are fastened with springs and dampeners.

Anecdote

The treadmill in the international part of the ISS was named after US comedian Stephen Colbert (*COLBERT*), an acronym for *Combined Operational Load Bearing External Resistance Treadmill*. Stephen Colbert initially won a competition to name a spacestation module. The US comedian asked everyone watching his TV show to vote for him. However, NASA decided not to call the entire spacestation module Colbert, a rather uncommon word for this purpose, but to name it *Tranquility* instead. As compensation, Colbert became the name of the new treadmill, and the name was invented to somehow match the acronym.

As a new exercise modality, research is currently conducted to test the efficacy of *human centrifugation*. A human centrifuge with a particularly short radius is used to apply centrifugal forces instead of real gravity. While spinning, exercise routines can be performed, such as heel raises or squats. It is also currently not clear whether one long trip per day or several short trips per day in the centrifuge deliver the best results, and which G-level is desirable. More on human centrifuges in Sect. 5.9.

5.5 Air and Pressure

On Earth, medical problems are known that are connected to changes in pressure from diving and mountaineering, while symptoms usually do not show up in daily life. Those who have not travelled to high altitude places like Bolivia or Nepal usually know little about the signs and symptoms of high altitude sickness, and have never experienced how sensitive they are to sudden exposure to lower atmospheric pressures. Humans are extremely vulnerable when it comes to decreases in air pressure. The level of susceptibility is independent of the endurance or power training state of an individual. Adaptation to altitude is an entirely separate entity and has nothing to do with how fast someone runs a marathon or how fit one is in general! In spaceflight, air pressure and air composition play very important roles for crew health. This is not only true for life on board spaceships and spacestations, but especially for spacewalks and the human presence on any celestial body other than Earth. Precisely speaking, the *partial pressure* of oxygen is the most important factor for human survival. The total pressure is the sum of the partial pressures of all gases in the air.

Decompression sickness (DCS) is the acute reaction to a sudden loss in pressure (for example, when ascending too quickly from a dive or as a result of a hole in the spacesuit), while *altitude sickness*, also called *acute mountain sickness* (AMS) comprises a different set of symptoms that occur when the pressure decreases a bit slower, for example, when hiking up a mountain in the Himalayas.

Decompression sickness, also called caisson disease or diver's disease, is a serious medical emergency! Sudden pressure loss leads to air bubble formation in the body, just like when opening a carbonated water bottle. The bubbles mainly consist of nitrogen that is usually dissolved in the body tissue and stays there due to the surrounding pressure. It behaves just like the air inside carbonated water while the bottle is still closed and the pressure kept constant. When this pressure decreases, gas bubbles form and collect in the muscles and joints, or travel through the blood stream to the lungs and brain. Air bubbles may obstruct blood vessels and stop blood from delivering oxygen to the organs. This is called gas embolism. When the oxygen supply of the brain is impaired, symptoms may be similar to a stroke and include cognitive deficits, paralysis, confusion and even unconsciousness—none of them very useful when on a space mission. Gas collection in the muscles and joints causes pain, called *the bends*.

Three severity levels of decompression sickness are distinguished:

- Type 1: *Bends* and itching,
- Type 2: Cognitive impairment and neurological symptoms,
- Type 3: Long-term disability due to non-reversible symptoms, including bone infarction (death of bone areas), hearing and vision problems, neurological impairment.

In spaceflight, the threat of decompression sickness is highest during spacewalks (EVAs), when either a micrometeoroid hits the spacesuit or the suit leaks for another reason (Sect. 4.4). As a prevention measure for pressure-related illness, spacefarers need to ideally spend the night before the EVA or at least a few hours in a pure oxygen atmosphere or perform exercises in the spacesuit. It makes the other gases including nitrogen follow the concentration gradient and leave the body. The nitrogen leaves the tissue and can therefore not form bubbles anymore in case of depressurisation. The procedure of sleeping in the *airlock* before the EVA is called *camping out* and also includes a slow decrease of the air pressure to approximately one third of the normal air pressure on Earth. This further reduces risks and also prevents the spacesuit from being inflated with too much force, making it easier to move. On the ISS, the *Quest Joint Airlock* is mainly used for this purpose (Fig. 5.22).

If symptoms of decompression sickness (such as bends) occur, the crew member needs to be brought to the airlock immediately. Inside, the air pressure has to be slowly increased while still keeping oxygen at 100%. The pressure can later only be released very slowly to prevent re-occurrence of symptoms.

On Earth, humans are mostly exposed to low air pressure when hiking or working in high mountains or high-flying aircraft. The higher the mountain, the lower the air pressure will be. Breathing frequency and heart rate increase to maintain sufficient oxygen supply to the body. If the oxygen supply is insufficient, this is called *hypoxia*. Gas bubbles in the blood or tissue do not occur in this scenario, as the drop in pressure takes place slowly and the nitrogen can be exhaled via the lungs. In hypoxia, the pulse rate is increased and moving feels a lot harder. The body uses more calories because the heart

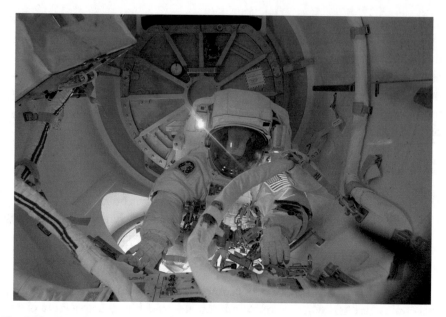

Fig. 5.22 American astronaut Jessica Meir exiting the Quest airlock for her spacewalk of 7 h and 17 min together with Christina Koch in 2019, exchanging a failed battery charge-discharge unit (Image: NASA)

beats faster, too. During longer stays at altitude, the body adapts by producing more red blood cells and providing more haemoglobin for oxygen transport. Hypoxia stimulates *erythropoietin* (EPO) production and thereby increases haemoglobin levels. Athletes use this effect during altitude training when exercising at an altitude of at least 2500 m. Some live high and train low, while others prefer to live high and train high. Altitude training increases speed, endurance, strength, and helps athletes recover faster. In a pressure chamber, it could also be used in spaceflight when preparing for strenuous operations.

When humans expose themselves to hypoxia too quickly while climbing a mountain (usually higher than 2500 m) too fast, they will get symptoms of *acute mountain sickness*. Symptoms include headaches, nausea, dizziness, sleeplessness and discomfort. In severe cases, swelling of the brain may even lead to confusion, unconsciousness and death (*High Altitude Cerebral Edema*, HACE). Shortness of breath, weakness and chest tightness may be signs of a *High Altitude Pulmonary Edema* (HAPE). In case some of these symptoms start to show up while ascending in the mountains, it is vital to descend as much as possible as quickly as possible. Altitude Medicine differentiates several altitude levels: high altitude (1500–3500 m), very high altitude (3500–5500 m) and extreme altitude (above 5500 m). See Fig. 5.23 for details. The atmospheric

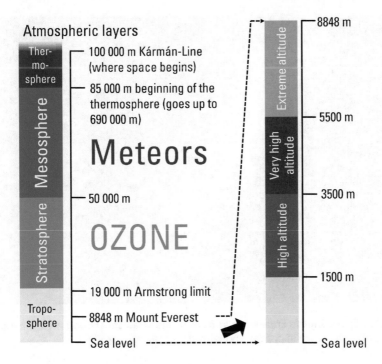

Fig. 5.23 Overview of the Earth's atmosphere and layers. Meteors appear in all atmospheric levels and may show up as shooting stars

layers can be seen on a photo of Space Shuttle Endeavour taken from the ISS, showing the atmospheric layers of the Earth (Fig. 5.24). During a mountain expedition, it is important to never sleep more than 500 m higher than the night before. It is recommended to first acclimatise at about 2000 m for at least a week, while carefully going on day trips to higher levels. The main reason why so many tourists do not make it to the top of for example, Mount Kenya is that some tour operators plan the night stays in huts that are further apart than these 500 m of altitude. Despite the best training shape and fitness level, people will not be able to make it to the top unless they are lucky and genetically better equipped than the average human. The same rules are true in spaceflight: in a spaceship or station, the air pressure should never be reduced quickly, but carefully and slowly.

The *Armstrong limit*, named after the U.S. Air Force surgeon Harry Armstrong, is the altitude above which water boils at the temperature of the human body (approximately 37 °C). This is the case at a pressure of 67 hPa and at an altitude of approximately 19,000 m. The pressure at sea level is around 1000 hPa where water boils at 100 °C. On the Earth's highest

Fig. 5.24 Space Shuttle Endeavour in front of the stratosphere and mesosphere. The troposphere shows up in orange and with clouds (Image: NASA)

mountain, Mount Everest, the air pressure is 314 hPa and water boils at 70 °C. At this altitude, most people have enormous problems acclimatising, and many need to breathe additional oxygen. The Armstrong limit is so relevant, as it is where body fluids start to boil. The lung cannot work anymore when the water in it evaporates, air bubbles enter the bloodstream, and this is usually lethal within a few minutes. This is what happens to someone who leaves a spaceship into space without wearing a pressure suit. Pressure suits should already be worn long before reaching the Armstrong limit. Fighter pilots usually start using them at altitudes of around 15,000 m. When breathing pure oxygen at 15,000 m, the oxygen partial pressure is equivalent to that of the normal atmosphere at an altitude of 4700 m.

> **Hint for Aspiring Spacefarers** To prepare for a spaceflight, the readers might want to get a feeling for the symptoms of hypoxia and try a fast ascend to high or very high altitude. This can be done by taking a cable car or gondola to a high mountain. The authors recommend for example the highest mountain of the Alps, Mont Blanc (4808 m). A gondola runs from Chamonix to the Aiguille du Midi (3842 m). Upon arrival, an increased heartbeat and decreased performance when climbing stairs or performing other types of exercise can be observed. A fitness tracker with integrated heart rate monitor may be used to get objective data. Similar experiences can be made in a pressure chamber. It is impressive to see how people in hypoxia become unable to solve the easiest maths equations without even realising it.

5.6 Further Medical Phenomena

Space Motion Sickness

The joy of finally having arrived in space and orbiting the Earth in a real spaceship is often dampened by *Space Motion Sickness* (SMS) as part of the *Space Adaptation Syndrome*. Symptoms usually persist for a few days, while the brain gets used to the new mismatch of information coming from the eyes and the vestibular system in the inner ear regarding positioning and orientation of the body. Nausea and vomiting occur similar to seasickness and the breathing frequency is elevated. Many experience problems with gas getting stuck in their bowels, and headache and orientation problems may add up to an unpleasant experience. Spacefarers may throw up unexpectedly and without warning. Smells and head movements usually worsen symptoms, and it is recommended to remain still while fixating an object with the eyes to calm things down. When entering weightlessness for the first time, it is best to not try flips in the air, but to stay calm and restrict movements to a minimum, while holding the head as fixed as possible. The more often someone has been to space, the less pronounced the symptoms will be. Freshmen spacefarers are usually hit worst. Luckily, symptoms are reported mild in most cases, but 10% of crew members are struck by more severe symptoms that affect their ability to work. Symptoms usually disappear within 2–4 days. NASA has tested the efficacy of *Promethazine* as a treatment, a medication that is also used against hallucinations and delusions. It seems to improve the symptoms of SMS. 60%

of spacefarers treated with it found it effective. The side-effect of sleepiness is usually compensated by the excitement of being in space. As throwing up in an EVA suit could be life-threatening, EVAs are only permitted after at least 72 h upon arrival in weightlessness. When making schedules for activities during the first days in space, decreased operational capability due to SMS should always be assumed. Soviet cosmonaut *Gherman Titov* (the second person in Earth orbit) in 1961 was the first to report symptoms of Space Motion Sickness in spaceflight history. He holds the record of being the first human who has thrown up in space.

Anecdote

In the 1950s and 1960s, NASA conducted experiments with deaf people who do not experience space motion sickness, as their inner ear is unable to sense and transmit the motion information to the brain. The deaf participants were able to enjoy parabolic flights, boat trips in the rough sea and the wildest experiments without even the slightest notion of discomfort.

The Eye

Spaceflight confronts the eyes with several problems at once. Elevated radiation levels increase the likelihood for a *cataract* (when the lens turns cloudy). Later in life, it might be necessary to replace the cloudy lens by an artificial one. Cataracts are also a problem in humans and animals on Earth, just the probability of occurrence and its severity increase the longer someone has been in space. In addition, flattening of the eye may lead to long-sightedness (hyperopia). That means, the person can see clearly in the distance but blurry in proximity. Another aspect of weightlessness is that dirt and dust do not fall to the ground, but float freely in the air. Particles tend to collect in the eyes and can be an annoyance. To limit this effect, a constant air flow usually carries the dust to filters to get it out of the way. On top of all these things, the *Spaceflight-Associated Neuro-Ocular Syndrome* (SANS), persisting vision problems after long-term spaceflight, is currently a major issue in research. The syndrome was initially called the *Visual Impairment and Intracranial Pressure* (VIIP) syndrome until 2017 and re-named to describe it more accurately. The same symptoms have been observed in 6° head-down tilt bedrest and were initially thought to be caused by increased pressures due to the fluid shift. The optic nerve swells and folds show up in the ocular fundus (back of the eye). More recently, however, scientists found out that the brain actually shrinks during spaceflight (Van Ombergen et al. 2018). These findings were confirmed by other groups. In addition, blood clots were found in the veins of the heads of astronauts potentially contributing to the vision problems. Current research indicates that artificial gravity exposure through human centrifugation might help to prevent SANS from occurring.

Infections and the Immune System

Simply speaking, in space the immune system is thrown out of whack! Its function is to protect the body against attacks from pathogens such as bacteria and viruses. It also detects damaged or mutated body cells, such as cancer cells, and destroys them before they do more harm. In space, many of these jobs cannot be performed properly due to immune system dysfunction. For this reason, more than half of Apollo and Skylab astronauts had bacterial or viral infections. The first research experiments on the immune system were conducted on spacestation *Skylab* in the 1970s. Reduced activation and activity of immune cells was discovered, especially of the white blood cells. Many experiments followed in subsequent years to find that the acquired (also called adaptive) immune system, that learns about new aggressors, adapts and memorises ways to fight them, stays more or less intact. At the same time, the innate immune system does not function in weightlessness. Its components mainly recognise foreign cells and invaders, attack and destroy them. These include cytokines, proteins that regulate cell growth or differentiation and immune globulins. Immune globulins are antibodies, produced by cells to act against specific aggressors. Also the function of white blood cells, for example, granulocytes, is inhibited in spaceflight. They are responsible for a general, non-specific immune response to pathogens. Natural killer cells are the ones to identify and destroy cancer cells (tumour cells) and cells infected by viruses. Their activity and efficacy are also reduced in weightlessness. Consequently, more tumour cells and infected cells can survive, develop and reproduce in the body. In addition, while the activity of the immune system is decreased, latent viral infections can re-activate. These infections are usually suppressed and the viruses only survive in niches such as nerve ganglia. Once the immune system malfunctions, these pathogens start to replicate and cause problems again. An example is chicken pox, that may lead to shingles (a very painful infection) when re-activated.

But how does weightlessness cause this? Why are these specific cells affected while others continue functioning as normal? Most likely the reason is the sense of orientation (up and down) that some cells rely on while others do not

(Braun et al. 2018). Particles inside the cells follow gravity and thereby indicate what is up and what is down. The cells' skeleton is constructed depending on this information. Without spatial orientation, the cell contents are built in a more chaotic way, leading to declines in cell function. In bedrest studies or other space analogues used to study the effects of spaceflight on the human body, immune system changes cannot be replicated. They can only be properly studied in real weightlessness.

Poisonous Substances

Spacecraft and spacestations contain many chemicals harmful to humans. These include ammonia and dinitrogen tetroxide. Ammonia (NH_3) is used as a cooling agent and may leak into the air inside space vehicles. Sometimes breaking news like "Ammonia-alarm: parts of the International Space Station evacuated" appear in the media, for example, in January 2015. Ammonia intoxication causes several symptoms. These include dizziness and sleepiness, coughing and eye-irritation, as well as neurological symptoms such as spasms and movement disorders. When inhaled, ammonia may lead to chemical burns of the mucous membranes, pulmonary oedema and even respiratory arrest. In addition, the liver may take serious harm and liver failure may occur. In case any of these symptoms show up while in space, one should always assume ammonia leakage and act accordingly.

Dinitrogen tetroxide (N_2O_4), an oxidiser used as rocket propellant that is typically applied in control thrusters, is highly poisonous, too. When humans get in contact with it, or with its combustion products serious chemical burns and pulmonary oedema appear. During the Apollo-Soyuz Test Project, combustion exhaust of 250 ppm NO_2 were sucked into the cabin (clearly exceeding the toxicity dose of 105 ppm), astronauts suffered from burning eyes, itching skin, chest constriction, burning chests and coughing. After landing, symptoms improved. Chest X-rays showed water in the lungs (pulmonary oedema). Luckily, the crew recovered quickly and entirely. Higher concentrations and a longer exposition time can lead to severe problems and death. It is most crucial to stop exposure.

Simultaneous appearance of headaches among several crew members may hint to increased carbon dioxide levels. A CO_2-concentration of 2% at sea-level air pressure causes headache in most individuals. On board spacestation Mir, headaches pointed out increased CO_2-levels twice, caused by faulty CO_2-scrubbers. Nowadays, portable CO_2-sensors indicate increased levels much earlier.

Another threat to crew health is carbon monoxide (CO), released from partial oxidation of carbon-containing compounds. It binds to the oxygen-transport-protein *haemoglobin* and inhibits oxygen transport in the body, leading to internal suffocation (internal because the lung still does a normal job, but the oxygen can no more be delivered to the cells). The first symptoms are headaches and a darker red skin. On board spaceships or spacestations, one should grab an oxygen mask as fast as possible and breathe pure oxygen. The further therapy of CO-intoxications may include treatment in a pressure chamber or airlock with pure oxygen to remove the CO from the haemoglobin. In 1997, several crew members on board Mir had symptoms that matched mild CO intoxication after an on board fire. They were observed for 24 h, but symptoms improved, so no further treatment was necessary.

Injuries

Smaller injuries are part of everyday life in space. When working in weight-lessness, people often hit objects or collide with one another. Haematoma (bruising) and small wounds such as abrasions are most frequent. Abrasions are primarily caused by Velcro used to attach objects. In case of deeper cuts in weightlessness, blood forms a sphere around the wound, instead of running down the skin. Apparently, this is easier to handle than on Earth.

Injuries of the foot or ankle on Earth are associated with swelling, bruising and pain. In weightlessness, swelling and bruising are much less pronounced, as the headward fluid shift counteracts these symptoms. Without the hydro-static pressure pumping blood into the injured area, the situation is similar to putting the leg up on Earth. It is important not to underestimate the severity of an injury in space, just because it often does not look as serious as it would on Earth. Two out of three treatments recommended on Earth already automatically take place in weightlessness: lifting the limb up and unloading. Another measure recommended is cooling with cool-packs.

During EVAs, objects with a great mass may be moved around, such as spacestation modules. Even though the heaviest objects may be shifted easily in weightlessness, crush injuries can occur through mass inertia. In addition, EVA suit gloves frequently cause pain and bruises under the fingernails that may be irritated and lifted up. Apparently, glove construction does currently not have a better alternative at hand. To prevent problems, it is recommended to cut fingernails short and tape them before going on an EVA.

Further injuries typical to spaceflight are strains and overuse problems, as well as rotator cuff (tendons in the shoulder) injuries. David Scott, for example, injured his rotator cuff during Apollo 15 while using a hand-drill on the surface of the Moon to obtain drill cores. Scott also suffered from haematoma under his fingernails (see Johnston et al. 1975). In addition, burns have been reported several times in spaceflight history. In 1997 during the Shuttle-Mir increment 4, an astronaut burned himself badly when a lithium perchlorate canister (an oxygen candle, see Sect. 2.5) caught fire, and a 1 m long flame hit his forearm. In space, burns also occur when skin is directly exposed to UV-sunlight. It is therefore essential to have UV-filters in all windows and visors.

When landing in a Soyuz capsule, injuries often occur during impact due to the confined space within the cabin and contact with objects. These injuries are usually minor or moderate and mainly located in the extremities and the back. Minor whiplash injuries have also been reported after landing. Seat belts and restraints often cause contusion.

Medical Incidents in Space

In spaceflight history, several medical events have been associated with urgent need for action and changes of mission plans. As mentioned in Sect. 5.1, serious cardiac arrhythmias appeared in Apollo 15. In 1987, a cosmonaut had to leave Mir earlier than scheduled when episodes of arrhythmias re-occurred time and again. On Soviet spacestation Salyut 7, in 1985 a cosmonaut suffered from a severe infection of the urinary tract that developed into a pyelonephritis (infection of the kidneys) and prostatitis (infection of the prostate). For this reason, the mission had to be terminated earlier than planned. Another more frequent issue is painful kidney stones. They may form when bone resorption is increased, as explained in Sect. 5.3. Toothache has also appeared on board spaceships and spacestations. In case of severe pain, a tooth might have to be pulled out, as an experienced dentist is usually not at hand. In spaceflight, more severe medical incidents requiring interventions like resuscitation, defibrillation or surgery are extremely unlikely.

In the early years of human spaceflight, several rockets exploded on the launch pads. These included Kosmos 133 in 1966 and incidents with Kosmos-3M and Vostok-2M rockets in 1973 and 1983. In 1967, cosmonaut Vladimir Komarov passed away when crash landing with Soyuz 1, when the main parachute did not open. In 1971, three cosmonauts suffocated on board Soyuz 11 when an air valve opened too early during reentry and the pressure suddenly dropped. The Space Shuttle programme had a big problem: two shuttles were destroyed in fatal accidents, killing everyone on board. In 1986, Space Shuttle *Challenger* exploded 73 s into launch and all seven crew members died. In

2003, Space Shuttle *Columbia* disintegrated upon reentry, and again none of the seven crew members made it. A setback for commercial spaceflight was the crash of the *SpaceShipTwo*-prototype *VSS Enterprise* in 2014, where only one of the two pilots was able to rescue himself with an ejection seat. Aspiring spacefarers should be aware of these risks and consider if the adventure outweighs them. Overall, none of these events stopped mankind from continuing with crewed human spaceflight.

5.7   Radiation

Earth is shielded from most radiation (all sorts of energies and particles) by its magnetic field and atmosphere (Fig. 2.3). When charged particles hit the Earth's magnetic field, they are either deflected or enter the atmosphere. In the atmosphere, they strike matter (atoms and molecules), react with them and cause showers of secondary particles. Only a tiny proportion of the *cosmic radiation* reaches sea level. Leaving the Earth's atmosphere and magnetic field, however, humans are exposed to the full amount of cosmic radiation: a serious health risk! As this topic is so highly relevant and its impact often neglected by those wanting to sell tickets to the Moon, Mars and elsewhere, it will be presented in more detail.

A few definitions are required to be able to put the impact of radiation on the human body in perspective. The unit *sievert* (Sv) is a SI unit to measure the radiation dose in body tissue (SI = International System of Units, which means the unit is international and official). Due to differences in composition, body tissues vary in sensitivity to radiation. The *background radiation* on Earth comprises around 3.5 mSv per year. An increase in cancer risk is traceable when someone has been exposed to a dose of 50 mSv and more. Acute radiation sickness is caused by 1 Sv per hour and more. At 5 Sv per hour, 50% of people exposed will die from acute radiation sickness. In the Fukushima reactor, doses up to 10 Sv per hour were measured during the Fukushima Daiichi nuclear disaster in 2011. This means that people directly exposed could not have survived these radiation levels. The fire fighters who had to work close to the reactor at the Chernobyl disaster in the Soviet Union in 1986 were exposed almost without protection to 16 Sv per hour, which meant almost certain death.

The ISS orbits at an altitude of approximately 400 km above Earth and is still somewhat protected by the Earth's magnetic field. Spacefarers who spend 6 months on board the ISS are usually exposed to 100–125 mSv, depending on solar activity. About 85% of radiation on the ISS is direct *cosmic radiation*, and 15% are secondary scattered particles from the spaceships hull (neutrons and protons). In an orbit around Mars, the radiation dose is 2.5 times higher than on board the ISS, as Mars lacks a magnetic field. During an EVA in Earth orbit, radiation levels are again higher compared to within a space vehicle, as radiation penetrates the EVA suit more easily. In further distance, the Earth is surrounded by the *Van Allen radiation belts* (Fig. 2.3), where some types of radiation (protons and electrons) are trapped and radiation levels are particularly high. Protons and electrons oscillate between the North and South Poles. The Van Allen belts were discovered in 1958 by the space probe *Explorer 1* that had a Geiger counter on board. Van Allen published the results in 1960, only a few years before the first humans went to the Moon, travelling through the radiation belts. An EVA within one of the Van Allen belts may expose a person to 0.4 Sv, while the annual maximal allowance for US astronauts is 0.5 Sv. Luckily, it is possible and recommended to plan flight routes in a way that the Van Allen belts are only crossed shortly within a few minutes. However, it is important to take the belts into account whenever planning a spaceflight.

Sources of Radiation

Earlier, the term *cosmic radiation* was mentioned without further explanation. But what is cosmic radiation? The term describes charged particles ever present space and moving at nearly the speed of light, including protons, helium and heavier nuclei and electrons. These particles cover a large energy spectrum, high energy protons occupying the top end.

The primary source of the lower-energy particles is the Sun. The solar wind comprises mainly protons and nuclear fusion products from the nuclei of the Sun. Processes in the solar corona, such as solar flares, coronal mass ejections and high speed streams, however, are only capable of supplying particles with a certain amount of energy. Particles with higher energies (approximately from one gigaelectronvolts, GeV) do not come from the Sun, but rather from outside the solar system. The next higher energy levels usually originate from stellar explosions (supernovae, Fig. 5.25) and possibly from collisions of neutron stars and black holes inside the Milky Way. The massive amount of energy released during these processes leads to a mix of high energy nuclei from all sorts of stable elements. At the upper end of the cosmic energy scale, cores of actives galaxies outside the Milky Way are considered the source. These are very dense and compact regions at the centres of galaxies. Further details, however, yet need to be resolved. The currently leading scientific model suggests large

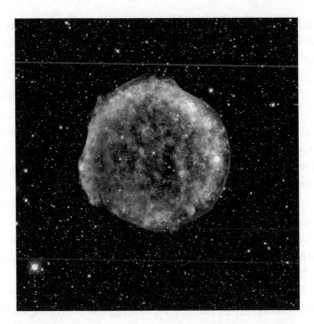

Fig. 5.25 Infrared and X-ray photo of the remnants of Tycho's Supernova that was visible on Earth in 1572. It was named after astronomer Tycho Brahe who observed and described it (Image: MPIA/NASA/Calar Alto Observatory)

black holes of many millions of times the mass of the Sun involving complex plasma processes that result in the ejection of these high energy particles.

While the Sun is a very variable source of charged particles, the density of high energy particles coming from sources far away seems to be constant over the years. The flux of cosmic particles is only decreased (up to 30%) for a few days after a solar mass ejection, as the shock wave it pushes while travelling further through the solar system deflects the incoming galactic cosmic rays and thereby reduces their amount. This effect is called *Forbush decrease*.

Primary cosmic radiation hits the Earth's atmosphere from all directions, interacting with oxygen and nitrogen atoms, and thereby creating secondary radiation in the atmosphere. Secondary particles mainly include neutrons, protons, pions and myons. Because of these reactions, the highest radiation doses are encountered at an altitude of approximately 20 km above sea level. The amount of energy reaching the surface also depends on the latitude. The Earth's electromagnetic field deflects the charged particles and collects them at the poles. Particle fluxes are highest close to the magnetic poles, where they cause the atmospheric gases to glow, resulting in aurora.

A special feature of cosmic radiation are so-called HZE ions = *high (H) atomic number (Z) and energy (E)*, coming from sources outside the solar system. This radiation type is especially dangerous for cells that do not divide, such as brain cells, since single-hit DNA damage can destroy the entire cell in one go or cause irreparable DNA damage possibly leading to cancer.

In addition to cosmic radiation consisting of charged particles, *gamma rays* (photons with high energy) are another potentially dangerous type of radiation. Even though they also come from cosmic sources, they are not counted as cosmic rays, as that term only refers to charged particles. Primary sources of gamma radiation are likewise supernova explosions and cores of active galaxies. Gamma rays cannot be deflected by magnetic fields, but a large fraction is shielded by the Earth's atmosphere. In spacecraft, shielding from gamma rays requires a thick layer of a dense material, like lead or tungsten— both are materials with a high mass and therefore almost impossible to use in spaceflight due to payload restrictions.

Figure 5.26 shows the penetration depths of different radiation types with regard to the human body. Roughly speaking, the following radiation types are relevant:

1. *Alpha particles* (helium nuclei): penetration of only a few micrometres, only affecting the skin.
2. *Beta particles* (electrons): penetration depth depends on energy: 0.5 cm per megaelectronvolt. Deeper organs may be affected.
3. *Gamma radiation*: penetration of all organs.

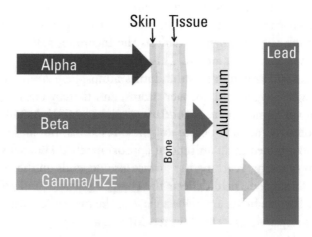

Fig. 5.26 Depiction of penetration depth of alpha-, beta- and gamma radiation

4. *HZE particles*: penetration of all organs, higher energies and more destruction than gamma radiation.

Effects of Radiation on the Human Body

Radiation causes damage to the body by either ionising atoms and molecules or by direct destruction (HZE). The energy delivered by radiation enables electrons to exit the atom or molecule and leave an ion with positive charge behind. These changes can induce chemical or biochemical reactions that may lead to alterations of the genetic material and the development of cancer. Due to differences in penetration depth, the individual types of radiation have varying target organs. Vulnerability of an organ or tissue depends on the cells' life expectancy or turnover, and on the amount of stem cells available. Stem cells create new cells to resupply the tissue. In case high energy particles of cosmic origin move through the body, energy absorption may start a series of reactions. Ionising radiation and secondary particles do not only cause direct damage, but also indirectly affect processes and structures, again leading to further problems. Creation of free radicals is an example. These atoms or molecules react intensely and may damage cells and cause cancer and other diseases with a longer delay. Overall, biological effects of ionising radiation induce health effects over a long time span, be it direct or by alterations in genetic material even in the next generations. When looking at the cell as such, damage to the enzymes, proteins and RNA molecules seem to cause less harm than damage to the DNA. The latter may include single or double strand breaks, defects or loss of bases or mistakes in the cross-linking of DNA base pairs. In addition, high energy particles may lead to chromosome damage. Chromosome fragments may get lost and wrong connections appear.

Every living organism has the ability to repair and compensate defects caused by radiation to a great degree. Damage to single DNA bases can be better tolerated and repaired than multiple defects at once or double strand breaks. Furthermore, mistakes may occur during repair when activating genes that were previously inactive. In the best case, this results in cell death and in the worst case, a tumour cell with uncontrollable replication and growth develops.

The higher the radiation dose, the higher is the probability for damage to the body. The likelihood to get cancer later in life especially increases with long-term spaceflight. Elevated mutation rates have been observed in the cells of astronauts. For airplane crew, the data situation is not clear. Due to relatively small numbers of spaceflight participants, data is limited regarding the cancer risks associated with spaceflight.

Acute Radiation Sickness

Intense radiation exposure over a short period of time and with a high radiation dose can lead to *acute radiation sickness*. Exposure from around 0.1 Sv may induce symptoms. Not only nuclear disasters such as the Hiroshima and Nagasaki nuclear bomb explosions or the Chernobyl and Fukushima nuclear power plant incidents, but also experience from radiation therapy have led to many cases and knowledge about symptoms. Haematopoetic (concerning the blood), gastrointestinal (concerning the gut and bowels) and CNS (central nervous system) symptoms are known and occur depending on the radiation dose. Latency can be observed between exposure and symptoms. It does not mean, if someone does not see or feel symptoms immediately after exposure that there is no problem! A dose of 0.1–1.0 Sv may lead to nausea, vomiting and fatigue. Some individuals do not observe any symptoms at all in this dose range. A reduction in white blood cells can be seen in the blood. Between 1 and 2 Sv, about half of individuals suffer from the symptoms previously described. In spaceflight, the maximally expected dose during solar coronal mass ejections and solar flares is in that range. If exposed to a dose of 2–3.5 Sv, the next symptoms to occur are fever, loss of appetite, diarrhoea and bleedings of the intestinal mucous membranes. Some people might not survive the consequences of exposure at this level of radiation. The higher the dose, the more will die in a shorter time span. Medical treatment is very demanding and involves administration of fluids, antibiotics, pain killers, blood transfusions, intravenous nutrition and many further components that would usually not be provided in a spaceship.

The actual radiation dose spacefarers would be exposed to on the way to Mars and its consequences are hard to estimate. Prediction of solar activity is difficult and knowledge about high energy particles on the way to Mars is limited. In addition, biological effects of high energetic radiation are hard to anticipate due to lack of data and experience. So far, only the Apollo astronauts have left the magnetic field of the Earth at a time of low solar activity.

Long-term damages after radiation exposure include a number of cancer types, such as leukaemia (blood cancer, usually shows up 7–15 years after exposure) and solid tumours, such as breast and lung cancer, tumours of the gastrointestinal tract and many more that may only appear years later. For this reason, space organisations consider sending older humans on missions to

Mars and other celestial bodies. Their idea is that the person is old enough to not lose too many healthy years by a potentially malignant disease.

Concepts for Radiation Protection in Spaceflight

When planning long-term human space missions, concepts for radiation protection and what to do in case of solar events are required. The following options are currently considered:

- Warning systems for *solar flares* and chambers (*storm shelters*) with special radiation isolation, such as a water layer, magnetic field or especially thick metal hull,
- Radiation vests that shield away parts of the radiation (such as the AstroRad vest made of high-density polyethylene),
- Freezing bone marrow for reimplantation after return,
- Cryogenic storage of sperm and ovarian tissue,
- Astronaut selection: individual genetic analysis and estimation of likelihood and predisposition for cancer,
- Supplementation of vitamins A, C, E and micronutrients (for example, selenium, zinc and copper).

It is important to measure the actual radiation exposure with dosimeters to be able to react to sudden increases. Radiation exposure should be kept to a minimum in the frame of what is possible. This concept is called *ALARA* (= *as low as reasonably achievable*). Thresholds for astronauts are usually at 3% REID (= *risk of exposure-induced death*), but that will be exceeded by far on trips to Mars with current technology. For this reason, it is even more important to build storm shelters and implement as many countermeasures as possible.

5.8   Medical Treatment on Board

Typical health effects of spaceflight and their causes were discussed in previous chapters. It is time to look into treatment options and further details. Questions such as *what to take along, what one wants to be able to treat for how long* and *which medical qualification levels crew needs to have* are important ethical considerations. Their extent was less critical in times of Roald Amundsen and Robert Scott leading expeditions to the South Pole in 1911 and 1912, as much less knowledge and equipment were available at that time. In addition, nobody followed the adventure of their heroes representing humankind in real time by public media. As space and payload are limited in future long-term missions, ethical considerations are important when making medical concepts.

First of all, options for the treatment of major emergencies and wounds need to be available without a question. Furthermore, medically trained crew, availability of equipment and medication, as well as the possibility to communicate with Earth and receive professional advice are the three main elements. To plan further, it is important to collect information on likelihood and severity of possible medical events, as well as the equipment needed for treatment of each of those events to make an informed decision on what one wants to be able to treat to which extent.

These are lists of the most common medical problems in spaceflight, be it on the ISS, in the Space Shuttle programme, on Skylab or Mir spacestation:

1. Space adaptation syndrome (nausea and vomiting),
2. Back pain,
3. Skin irritation and dermatitis,
4. Irritation of the eyes, dry eyes, foreign bodies in the eyes,
5. Respiratory tract infection,
6. Urinary tract infection,
7. Cardiac arrhythmias,
8. Diarrhoea,
9. Headaches,
10. Decompression sickness,
11. Allergic reactions.

The most frequent injuries on board the ISS are:

1. Haematoma (bruises),
2. Blood and pain under the finger nails (Problem after EVAs),
3. Muscle strain,
4. Superficial wounds.

The most common symptoms are:

1. Loss of appetite,
2. Fatigue,
3. Sleeping problems,
4. Constipation,
5. Facial swelling,
6. Dehydration.

More severe and potentially lethal diseases such as appendicitis or gall bladder infection are not mentioned in the lists, because they are rare. However, in case they occur, they would probably be deadly without proper treatment. For a good decision of what to take, not only the most likely conditions, but also more rare but potentially very deadly ones need to be taken into account. Effective and lightweight treatment options that significantly decrease the probability of death need to be identified. On a space expedition, treatment of any possible incident can never be achieved 100%, as this usually requires heavy hardware that may only be used to treat a small number of problems. Lightweight technology and medications with a broad spectrum of effective indications are what is needed. As an example, antibiotics with a broad spectrum are required, capable of attacking a great number of bacteria, other than antibiotics that only work on a small number of pathogens. Another issue is to allocate as much payload as reasonably possible to medical equipment and medications. In the near future, it will not be possible to take computed tomography or magnetic resonance imaging devices on board spacecraft, but a simple X-ray tool and ultrasound should be considered. The capability to perform surgical procedures will always be limited by the level of training of the crew.

Every spacefarer needs to be aware of the limitations in medical treatment capability and know that it will be rudimentary compared to a hospital. The situation is probably comparable to going on an expedition in a lonely area of the Himalayas or Antarctica.

Crew with Medical Education

For longer flights, it could be considered to take one or two experienced and skilled medical doctors along. Training needs to include space and emergency medicine, trauma surgery, general medicine and dentistry. The person needs many years of experience in these fields and preferably on expeditions. In addition, some of the other crew members should also have at least experience, for example, as a paramedic or nurse to be able to react and assist. It is currently common practice to train selected crew members as *Crew Medical Officer* (CMO). Two crew members on board the ISS are required to have this training of 40 h that can be extended to 70 h if the astronaut agrees (*Space Emergency Training*). Even 70 h seem to be alarmingly short for flights outside the Earth orbit! Once leaving it and the option to always return within a few hours vanishes, much more extensive medical skills are desirable. Spaceflight has seen many medical doctors on missions.

Medical Care on the ISS

On board the ISS, medical tools, consumables, and medications are arranged in kits. These are the Convenience Medications Pack, Oral Medications Pack, Topical and Injectable Medications Pack, Medical Diagnostics Pack, Minor Treatment Pack, Medical Supply Pack, IV Supply Pack, Physicians Equipment Pack, and Emergency Medical Treatment Pack. Those medications most frequently used are provided in the Convenience Medications Pack. Items such as electronic equipment for medical purposes are contained in the Medical Diagnostics Pack. The Minor Treatment Pack contains dental and surgical tools, as well as consumables required for small surgeries. Further supplies such as syringes, bandages, and gauze are provided with the Medical Supply Pack. In addition, a portable clinical blood analyzer and a defibrillator are available on board. Checklists provide the information what can be found where. Furthermore, a *Human Research Facility* (HRF) provides equipment for basic human research. In addition, the *Environmental Health System* (EHS) monitors the environment on board (water quality, radiation, etc.). Also the exercise devices belong to the HMS (Sect. 5.4).

On the ISS, the crew is frequently required to take blood for scientific purposes. Figure 5.27 shows the procedure. Many further devices are available for all sorts of medical and physiological experiments, including eye testing (Fig. 5.28).

Surgery in Spaceflight

In weightlessness, dirt particles, microorganisms and tools float around and are a serious problem for surgical procedures. For this reason, technical solutions have been invented that enable for surgeries in spaceflight. Due to the fact that return from the ISS is always possible within a few hours, however, surgical infrastructure and tools have never been tested in flight, but only on the ground and in parabolic flights that offer weightlessness for up to 30 s. So how could surgery work out on board a spaceship?

First of all, having the skill to perform the surgery is essential. Weightlessness just makes it harder. The patient and instruments somehow need to be fixed, for example, by using the Crew Medical Restraint System. A sterile environment needs to be created by using disinfection agents and sterile materials and instruments. Contamination through the air deserves a lot of attention, not only because of the dirt floating around, but also because of the impaired immune system, that increases the likelihood of infections. Several

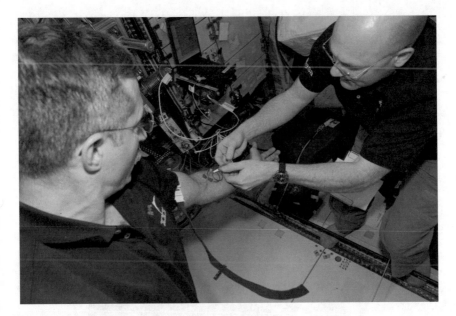

Fig. 5.27   Drawing blood on the ISS (Image: NASA)

design ideas for surgical chambers have been presented, usually applying *Laminar Air Flow* and a *glovebox* design. Laminar Air Flow shifts the air in one direction, preferably sterile air towards the wound, while suction collects dirt particles in a filter. Smaller surgeries, such as wound closures, certainly do not require such a big setup, but more invasive operations with greater wound surfaces might do so. The application of antibiotics as a prevention measure is recommended and also practiced in surgeries on Earth. When designing a medical concept for a longer space mission with more payload availability, the question needs to be answered whether or not to take equipment for minimally invasive surgeries. These allow for much smaller skin incisions. Anaesthesia should not be provided by gas, but using medications instead. The reason is the small air volume of a spacestation or spaceship that might otherwise lead to significant concentrations of the narcotic gas in the breathing air. In the 1980s, NASA planned a spacestation with the name *Freedom*, that was to include a sickbay module to test medical systems and develop such capabilities. The spacestation, however, was unfortunately never built. At some stage, a medical module was also planned for spacestation Mir. As none of these plans have become a reality, experience with medical and surgical treatment capabilities in spaceflight is extremely limited.

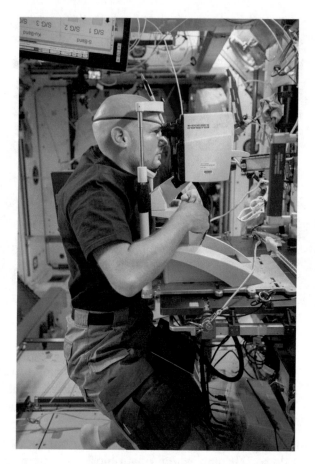

Fig. 5.28 German astronaut Alexander Gerst during an eye examination on board the ISS (Image: NASA)

Medications

There are several issues with medications in spaceflight that need to be considered when preparing a mission. Radiation destroys the chemical structure of medications and makes them less effective over time. That means, someone will need several tablets of the same medication towards the end of a mission, while one tablet would have been enough at the beginning. It is currently unknown which types of medications are affected to which degree, but it is certain that some drugs might lose a great percentage of their effectiveness. In addition, physiological changes in spaceflight, such as the fluid shift, might affect the efficacy of medications and how quickly the body is able to eliminate the degradation products. As it is unknown how much a medication is affected, in some cases, it might help to start with a normal dose and increase the amount in case the desired effect does not show up. This approach is called titration and may be used especially for pain and sleep medication. Due to losses in effectiveness, alternative treatment options such as postures, physiotherapy, cooling or heat should be taken into account and used as much as possible. In the Apollo programme, each medication to be used on board was tested on each astronaut beforehand to make sure allergic reactions do not endanger mission success.

Substance Abuse in Spaceflight

Performance-enhancing drugs with significant side-effects were very common in the early years of spaceflight. As failure was not an option, astronaut performance was improved by drugs that helped during the flight, but put the spacefarers' health at risk. In Gemini and Apollo, D-Amphetamine was part of the medical kit—a central nervous system stimulant (Berry 1975). It was even part of the lunar lander's kit. Despite its effects of making euphoric and increasing athletic (muscle strength) and mental abilities (shorter reaction time, better fatigue resistance), it causes addiction and may even lead to psychosis. The authors recommend not taking any risks by doping in spaceflight, as it might end up in psychiatric problems seriously endangering mission success.

Telemedicine

The first implementations of telemedicine date back to the 1930s, when several military branches, including the German Navy, started to implement medical consultations via radio. Telemedicine is nowadays well established and plays significant roles in many countries of the world. Rural medical care incorporates telemedicine structures at great scale. As an example, *the Royal Flying Doctor Service of Australia* (RFDS) provides medical care to the most remote rural places by radio and video communication. The farms are supplied with medical equipment and medication boxes that inhabitants are instructed to open and apply, if needed. In spaceflight, each crew member has a *crew surgeon*, a personal medical doctor who provides video conferences to discuss possible issues on a regular basis. NASA has set new standards in astronauts medical monitoring already in the 1960s and now uses *Smart*

Medical Systems to identify problems early. The development in this field is rapid. In possible future trips from the Earth to very distant objects, delays in communication will complicate telemedicine. In the worst case, a signal will require 18 min from Mars to the Earth, which means it would take at least 36 min until an answer arrives. Remote-controlled surgeries, for example, would be impossible. For this reason, medical self-sufficiency is of highest importance for these missions. In future private or commercial spaceflights, complete 24 h surveillance of crew and passengers does not seem necessary, but help in case of emergencies and medical problems needs to be provided.

5.9   Medical Research in Spaceflight

In spaceflight, medical research serves two objectives. The first is basic science with the aim to use weightlessness and isolation to answer fundamental questions regarding the human body. The second is the so-called operational research that aims to improve medical capabilities on space missions. These include the treatment of injuries and illnesses and the improvement of operational capabilities through better sleep, the reduction of pain and discomfort, exercise and prevention measures. The research is conducted not only in spaceflight, but also in analogue models, such as *parabolic flights, bedrest studies* and in simulated hypergravity in *human centrifuges*. These environments will now be discussed in detail. Space organisations define their scientific priorities in the so-called roadmaps. In the field of Medicine, the European Space Agency states in its *Roadmaps for Future Research* the priority on research dealing with

1. The impact of gravity on biological processes, cells and organisms,
2. Supporting life in hostile environments,
3. Understanding and preventing physiological adaptations to reduced gravity,
4. Psychological and neurosensory adaptations to reduced gravity, isolation and confinement, and
5. The cosmic radiation risk for human exploration of the solar system.

Since the 1970s, medical research has been a core element of human spaceflight. When it had been established in the 1960s that humans can exist and survive in the space environment, physiological changes and interventions to prevent problems moved into the focus of interest. Since that time, the space agencies involved scientists and medical doctors to design, propose and run medical experiments in flight. Such an offer to propose research by submitting a *proposal* is called an *announcement of opportunity*, a *call* or *solicitation*. In Europe, ISS experiments can be proposed approximately once every 5 years by all scientists. This way, scientists get the opportunity to have experiments in space without actually being employed by a space agency. *Flight experiments* include research to be conducted in spaceflight, while *pre-mission and post-mission studies* are performed before and after a mission. When designing such experiments, the small number of available subjects needs to be considered. Scientific articles often only include results of five or even fewer spacefarers.

In spaceships and spacestations, experimental devices are usually contained in so-called *racks* (*International Standard Payload Rack*, ISPR). These are kind of built-in closets in standard size (Fig. 5.29). When designing an experiment as a scientist, it is possible to either use the equipment available on board or apply for new hardware to be sent up (so-called *experiment unique equipment*). The more resources and crew-time an experiment takes, the less likely it will be accepted and implemented. Currently, the available racks for medical research on board the ISS are the *Human Research Facility* (HRF) (Fig. 5.30), the *European Physiology Module* (EPM), the *Muscle Atrophy Research and Exercise System* (MARES) and the *JAXA Onboard Diagnostic Kit* (ODK) in the Kibo Module. Medical research facilities on board include among other items:

1. A *Continuous Blood Pressure Device*, (CBPD),
2. An ECG-device that allows recordings for up to 24 h (Holter monitor),
3. The *Pulmonary Function System* (PFS) for lung research,
4. An ultrasound device,
5. A scale, called the *Space Linear Acceleration Mass Measurement Device* (SLAMMD),
6. A centrifuge for samples, called the *Refrigerated Centrifuge* (RC),
7. Equipment to collect blood, saliva and urine samples,
8. Activity monitoring devices (Actiwatch).

Bedrest Studies

Bedrest studies are a method to test interventions against the negative effects of immobilisation on a greater number of participants on the ground. As only very few individuals actually participate in real spaceflight, bedrest studies enable research that would otherwise take decades. Countermeasures can be tested without the need to transport the devices required to an actual spacestation. A large proportion of scientific publications from Space Medicine

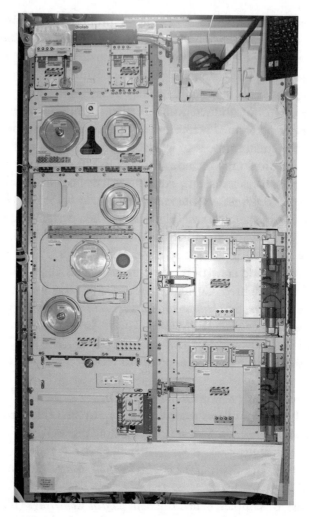

Fig. 5.29 The *BioLab*, a rack in the Columbus Module of the ISS. It serves for research of plants, cells and little organisms and contains, for example, an incubator, two centrifuges, a microscope and a spectrometer (Image: NASA)

are from studies using the bedrest model. Many effects of weightlessness can be reproduced when putting participants to bed with 6° head-down tilt. Bedrest studies date back to the 1950s and were introduced to make predictions on the effects of human spaceflight before the first humans went to space. It later became evident that bedrest is a great model for some organ systems including the musculoskeletal and the cardiovascular system. For other organ systems, such as the sense of balance or the immune system, real microgravity is needed.

Fig. 5.30 The *Human Research Facility* (HRF) in the ISS's Columbus Module consists of two racks. They contain the *Pulmonary Function System* (PFS) to test respiratory and cardiovascular function. Rack two is shown here (Image: NASA)

Many results from bedrest studies can be directly transferred to immobilised patients and are therefore highly valuable for the general population.

In any scientific study dealing with humans and their data, strict ethical rules apply. Prior to the study, an official ethical board needs to be consulted and their suggestions implemented. The subjects need to sign informed consent forms prior to being enrolled into the study. To make sure results of bedrest studies are comparable, studies are standardised according to standardisation plans. Standardisation plans usually agree on the following items (among others):

- Short-term bedrest lasts 5 days, mid-term bedrest 21 days and long-term bedrest 60 days.
- Studies may be conducted in a cross-over design. This means that all subjects participate in two bedrest campaigns, once with the tested intervention and once without. Half of the subjects receive the countermeasure in campaign 1, the other half in campaign 2: division of the participants into groups is randomised. This procedure is necessary to eliminate possible campaign effects influenced, for example, by the season or remaining changes induced by campaign 1 that are still affecting campaign 2.
- A set of standard measurements before and after the study guarantees quality control (tilt table tests, spiroergometry, bone density, etc.).
- Nutrition is standardised and adapted to body weight.
- Participant health is monitored by regular blood tests and daily physician ward rounds.
- A *Head Medical Doctor* has the responsibility to protect the participants' health. The person in this function may cancel and stop experiments in case of health risks.

Bedrest studies are usually conducted either in hospitals or specifically designed facilities. As an example, the German Aerospace Center (DLR) in Cologne, Germany, has a facility called "Environmental Habitat (EnviHab)", where bedrest studies can be run with 12 participants at a time. Bedrest is performed in 6° head-down tilt, as this mimics the effects of weightlessness best, especially the fluid shift. Participants are required to touch the mattress with at least one shoulder at all times. They are not allowed to stand up at all during the bedrest period! Even going to the toilet once daily would ruin the scientific results, as they would no more be comparable to spaceflight. Going for a number one and a number two works just like in a hospital using a urine bottle or bedpan. A shower stretcher is used to take a shower in 6° head-down tilt. Participants are allowed to read, play video games, watch movies, surf the internet and much more. Video surveillance is used to guarantee compliance.

Baseline data is collected prior to bedrest. Experiments differ between studies, so the timetable varies. As experiments sometimes take several hours and not all subjects can complete the same experiments in one day, participants usually arrive in pairs of two. This means that subjects A and B go through the first experiments on day 1, while participants C and D do the same experiments on day 2 and E and F on day 3 of the study. Due to circadian rhythms, it is important to roughly do the experiments always at the same time to guarantee comparability. The reason is that many hormones and the performance show fluctuations during the day. Bedrest starts a few days into the study when the baseline measurements are completed. Prior to this, participants are still allowed to walk around in the facility. The same is true after completion of the bedrest period, when participants stay in the facility for the post-measurements. In longer studies, rehabilitation protocols can also be part of the scientific testing and may be applied and compared.

Just as in real spaceflight, participants of bedrest studies need to be selected very carefully. Even though it appears to be an easy job to stay in bed this long, it is actually hard work and requires a lot of mental strength. The selection process usually comprises several steps, starting with a telephone interview to check the inclusion and exclusion criteria. It is very important to make sure not to harm participants with the study. This means that health risks need to be assessed, including, for example, a tendency for venous thrombosis. Some people have a genetically determined higher probability for blood clotting in the veins and need to be excluded. Other reasons for exclusion may be a decreased bone mineral density or that a blood value of importance to the study deviates. Most people are usually excluded from participation for various reasons. Regarding the personality, the desired character traits include a true interest in the scientific research, social competence and the ability to be a team player. The goal is to have as few drop-outs as possible to guarantee sufficient data and a satisfactory experience for the volunteers.

Bedrest studies are almost as strenuous as real stays in space, and there are often many reasons for people to lose their enthusiasm about the study. That is why a great atmosphere with many surprises and events is of highest importance. The following items are sometimes described as unpleasant:

1. Just like in spaceflight, back pain may develop.
2. The fluid shift towards the head may cause headaches.
3. Participants may experience homesickness.
4. Most subjects do not complete the work they had planned to accomplish during the study.
5. Participants often find it hard to eat all the healthy food provided (such as a lot of salad).

On the other hand, participation in a bedrest study is a great opportunity to experience something very unique, to learn about research, talk to scientists, have a break and make many new friends. Announcements of opportunities for bedrest studies and other research of the European Space Agency can be found on the website linked in Fig. 5.31.

Fig. 5.31 ESA website with their announcements of opportunity http://www.esa.int/ Our_Activities/Human_Spaceflight/Research/Research_Announcements

Parabolic Flights

Parabolic flights are the only possibility to conduct human experiments in real weightlessness without flying to space. An airplane flies a steep upwards ascent and then decreases engine thrust to enter a parabolic trajectory (descent). After a dive of approximately 22 s (phase of microgravity = weightlessness), the pilot pulls the stick back and increases engine thrust to end the free fall (Fig. 5.32). During acceleration and in ascent, a pull of approximately two G prevails. With this technique, weightlessness can be maintained for 22–30 s in each parabola, depending on the type of aircraft. For this reason, only short-term reactions to weightlessness may be observed and studied, such as blood pressure changes. Long-term effects of weightlessness may not be addressed in parabolic flights. In parabolic flight campaigns, it is common to fly up to 31 parabolas: one as a test and 30 for the actual measurements. Parabolas are usually flown in bouts of five, followed by a break. Parabolic flights can also be used to simulate the gravity on other planets and moons, such as our Moon and Mars.

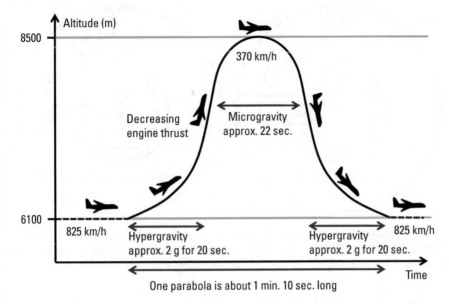

Fig. 5.32 Illustration of a parabola in a parabolic flight. *Hypergravity* and *microgravity* alternate

Anecdote

The airplanes used for parabolic flights are called *vomit comets*, as they give many participants a hard time. Only few individuals can go through a parabolic flight without nausea and vomiting. Medications are given to decrease symptoms (for example, scopolamine). When taking the plunge of a parabolic flight, one needs to consider that nausea, once started, will usually persist for the remaining parabolas that may become a long odyssey. Experienced parabolic flight participants recommend to eat croissants with marmalade for breakfast before the flight, as this is the least unpleasant on the way back out again ...

Human Centrifugation

Not only weightlessness is relevant to crew health during spaceflight, but also *hypergravity*, especially during launch and landing. For research and training purposes, hypergravity can be generated in human centrifuges on the ground. The principle is to spin the participant in a circle to generate *centrifugal forces*. Depending on the radius of the centrifuge and the rotational speed, these

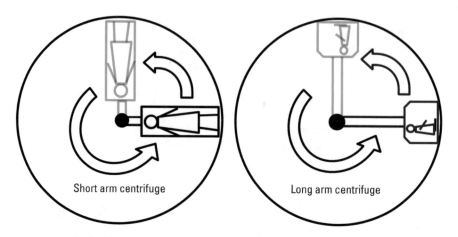

Fig. 5.33 Comparison of a long and a short arm human centrifuge. In short arm human centrifuges, the head is located close to the centre of rotation, while the feet are spinning much faster, resulting in a G-gradient within the body

forces may be very high. For this reason, human centrifugation needs to be conducted in well-designed devices and under supervision of a medical team. It is possible to simulate launch and landing profiles of various spacecraft and rockets, to train G-resistance and do research. There are two types of human centrifuges: long and short arm human centrifuges (Fig. 5.33).

Long arm human centrifuges have a much larger radius than short arm human centrifuges. The radius is often 5–10 m, while the participant sits in a small booth at the end of the arm. An airplane or spaceship cockpit may be copied to make the experience as realistic as possible. In a long arm centrifuge, the entire body experiences about the same G level, as the centre of rotation is several metres away. Long arm centrifuges are the common type and have been used in pilot training for many decades. Centrifuge exercises improve G-tolerance and enable fighter pilots to fly more extreme manoeuvres with higher G-forces. Long arm human centrifuges have also been used to simulate launch and landing phases of spaceflights for many years.

Analogous to this, one could build spaceships with rotating rings to generate artificial gravity and allow for a more healthy and normal life on board. Already early science-fiction movies such as *2001: A Space Odyssey* by Stanley Kubrick from 1968 suggested such options. However, as discussed in Sect. 2.2, it is extremely complicated to build such structures from an engineering point of view. While searching for alternative ways to apply artificial gravity in spaceflight as a countermeasure, the concept of short arm human centrifuges was developed. These have a short radius and may fit inside a spacestation or spaceship module. The participant's head is located close to the centre of rotation. Exercising against the centrifugal forces while spinning in a short arm human centrifuge seems to be very promising. In terms of physics, these centrifuges differ from a rotating ring-spaceship in two ways:

1. Head movements cause *Coriolis forces* by stimulating the canals of the inner ears in a confusing manner, leading to nausea and dizziness.
2. A G-gradient with lower G-forces at the head and higher forces at the feet leads to potentially different physiological reactions.

To study the value of short arm human centrifugation as a countermeasure to the effects of weightlessness on the human body, studies are currently under way. These will answer the question how long the intervention needs to be used per day, if it is better to apply it in one bout or in several, and which exercises are efficient. The next level will be to test a short arm human centrifuge in spaceflight.

Isolation Studies

Isolation studies are a great way to study the adaptation, physiology and psychology of humans in defined and controlled settings. As isolation is a very important aspect of spaceflight, these studies are a major component of research and intervention testing for long-term space missions. Isolation mainly causes social and psychological issues. The main question is which interventions help to decrease social tension and conflicts, as well as to improve mood, stability and wellbeing.

Among the more important isolation projects in history were the *Biosphere 2* experiments, the *Mars 500* study and *NASA's Extreme Environment Mission Operations* (NEEMO) under water station. Over-wintering in Antarctica is another suitable setting to study humans in isolation. Antarctica is of particular interest, as participants are completely isolated in a hostile environment without the option of rescue for several months. Therefore, the overall situation is more comparable to long-duration spaceflight than underwater isolation.

The *Biosphere 2* complex near Tucson in Arizona, USA, was built between 1987 and 1991, and contains several ecosystems with a variety of plants and a salt water lake (called ocean) under glass roofs. It encloses 1.27 hectares of land, providing a jungle, mangrove wetlands, savannah, desert and an ocean biome that originally contained corals. In addition, agricultural space and human housing complete the facility. In principle, the Biosphere is a large greenhouse with an atmosphere that can be isolated entirely from the outside, meaning the Earth's atmosphere. While the Earth's atmosphere is thought to be the first biosphere, the name "Biosphere 2" reflects its atmosphere's airtight separation. Two large *lung* buildings compensate the air pressure within the complex, especially when the air expands while the Sun shines and heats it up. The interior volume of the complex is connected to the lungs via tunnels. When the air pressure inside the system increases, a heavy lead plate sealed with thick rubber foil around it is elevated by the pressure within the lung building, increasing the total volume of air within the closed system. The air is circulated and cooled within the facility by large air conditioning machines located in the basement underneath the Biosphere. In total, two closed-system experiments were conducted: a long experiment from 1991 to 1993 and another shorter run in 1994. The first experiment included eight participants of both sexes. They lived in the closed system for 2 years and only ate what they produced. Even though the idea was to never add oxygen from the outside, nor remove CO_2, and though it was not directly made public, CO_2 removal became vital when the CO_2 concentrations reached dangerous levels, and oxygen was added as well. Water was recycled within the system. Electrical power was added from external sources. Even though media often reported that the experiment had failed, it was scientifically a great success. The fact that oxygen and CO_2 concentrations developed into such big problems, the growth of insect populations and bacteria, as well as the dying fish and coral population gave great insights into the challenges to be solved in future experiments. In fact, far more cubic metres of plants are needed and much less concrete can be tolerated by such a system. In addition, the desert biome hardly contributed anything to the oxygen production. The ratio of oxygen generation and consumption needs to be addressed by including many more plants, filling up as much volume as possible, while critically considering the introduction of oxygen consumers. As an example, monkeys lived in the Biosphere 2-complex during the first experiment, that did not contribute to the human survival, just consumed food and oxygen. The participants of the first campaign were hardly able to survive from the food they produced and all of them lost a lot of weight. A second study of 6 months was conducted in 1994, where food production was optimised and enough food was available. So far, another experiment of this type and size has not taken place anywhere in the world. In case any of the readers consider building a human outpost, the authors recommend visiting the Biosphere 2 complex and taking as many of the tours they offer, as possible. Biosphere 2 nowadays belongs to the University of Arizona (Fig. 5.34 and QR-Code in Fig. 5.35). One of the

Fig. 5.34 Interior view of the *Biosphere 2* near Tucson in Arizona, showing the mangrove wetlands and the ocean (Image: Colin Marquardt / Public domain)

Fig. 5.35 Website of the *Biosphere 2* http://biosphere2.org/

Biospherians, Mark Nelson, wrote a recommendable book about his time in the Biosphere 2 (Nelson 2018).

Mars 500 was a joint isolation study of the Russian Space Agency Roscosmos and the European Space Agency. Six male participants were confined in a simulated space mission for 520 days between June 2010 and November 2011. The NEK-facility is located in the IBMP (*Institute for Biomedical Problems*) in Moscow. The facility has four modules, 243 m^2 in total, including a room of 39 m^2 resembling the surface of the planet Mars or the Moon. The living module comprises 72 m^2. A medicine- and research module and storage space for food and consumables are also components of the facility. The simulation of a Mars mission included the descent in a spaceship and EVAs on the Mars surface inside the Mars room. Many experiments and routine tasks were completed by the crew to make the spaceflight experience as realistic as possible. Contact to the outside world was only possible by email and radio, and a delay in communication increasing with distance was simulated. Research mainly included physiological research, and results showed huge losses in muscle mass and fitness despite moderate exercise. Another aim was to prevent social conflicts by careful subject selection and previous psychosocial training. The experiment was completed successfully. At the moment (as of 2020), a series of shorter isolation experiments is conducted in the NEK-facility, called the SIRIUS Moon Project.

NASA's *Aquarius Reef Base* is an underwater research lab of the NASA NEEMO programme (Fig. 5.36). It is located 9 km off Key Largo in Florida in a depth of 19 m next to a coral reef. Built in 1986, it has now been used for more than 20 isolation missions, often including astronauts and scientists. The facility has space for four participants and two technicians. Medical as well as social and psychological research questions may be addressed, and

Fig. 5.36 Crew of NEEMO mission 16 next to the *Aquarius* (Image: NASA)

dangerous situations can be simulated using remote control of systems and video surveillance.

Several research bases are located on the continent *Antarctica*, where crews overwinter each year. This means, they stay in Antarctica during the southern hemisphere winter that is extremely dark and cold. The Sun does not rise for several months, while temperatures drop to around −80°C. The lowest temperature ever measured in Antarctica was −89.2°C (1989, Soviet Vostok Station). Due to the darkness and extreme climate, most Antarctic stations cannot be reached for many months, which means the inhabitants are entirely isolated. There are many parallels to spaceflight. Recently, some participants have completed questionnaires that showed they felt even more isolated than astronauts, while MRI studies showed a shrinking of the brain during isolation. The largest Antarctic outpost is McMurdo Station of the USA, capable of supplying more than 1200 people in 85 buildings. A 3 km long road connects McMurdo Station to the *Scott Base*, belonging to New Zealand. From 1962 to 1972, McMurdo station had a nuclear power plant. It was later dismantled and brought back to the USA. Especially the smaller stations can be considered good models for spaceflight due to the lack of diversity in social interaction and smaller facilities. Germany's Neumayer Station is one of these smaller stations built on the shelf ice almost opposite McMurdo station on the other side of Antarctica. Nine persons usually stay over the winter, which is quite comparable to a crew in spaceflight. Italy and France run Concordia Station at an altitude of 3233 m, offering a combined opportunity of isolation research in hypoxia. The European Space Agency solicits announcements of opportunity for research experiments on Concordia station. These include mainly medical and psychological research. Job opportunities for the stations in Antarctica are usually made public, and applications can be submitted by anyone meeting the criteria.

This is the end of the chapter on Space Medicine. For those readers who want to continue reading, the following books are recommended:

Gilles Clement has published the book "Fundamentals of Space Medicine", presenting space life science research in spaceflight in more detail (Clement 2011). The book "Wilderness Medicine" by Paul S. Auerbach comes with a great amount of medical knowledge in much detail, covering all possible aspects of wilderness and expedition medicine (Auerbach 2007). In addition, the most detailed and high-level specialist book available on Space Medicine was written by flight surgeon and astronaut Michael Barratt together with Ellen Baker and Sam Pool (Barratt et al. 2020).

References

Auerbach, P. S. (2007). *Wilderness medicine* (5th ed.) Philadelphia: Mosby Elsevier. ISBN:978-0-323-03228-5

Barratt, M. R., Baker, E., & Pool, S. L. (2020). *Principles of clinical medicine for space flight* (2nd ed.). New York: Springer. ISBN:978-0-387-68164-1

Berry, C. A. (1975). Medical care of space crews (medical care, equipment, and prophylaxis). In J. M. Talbot & A. M. Genin (Eds.), *Space medicine and biotechnology* (Vol. 3, pp. 345–371). Washington: NASA Scientific and Technical Information Office. NASA SP-374. In M. Calvin & O. G. Gazenko (Series Eds.), *Foundations of space biology and medicine.*

Braun, M., Böhmer, M., Häder, D. P., Hemmersbach, R., & Palme, K. (2018). *Gravitational biology I: Gravity sensing and graviorientation in microorganisms and plants. SpringerBriefs in space life sciences book 1.* Cham: Springer. ISBN 978-331-993893-6.

Buckey, J. C., Gaffney, F. A., Lane, L. D., Levine, B. D., Watenpaugh, D. E., & Blomqvist, C. G. (1993). Central venous pressure in space. *The New England Journal of Medicine, 328*(25), 1853–1854.

Buckey, J. C. Jr, Gaffney, F. A., Lane, L. D., Levine, B. D., Watenpaugh, D. E., Wright, S. J., et al. (1996). Central venous pressure in space. *Journal of Applied Physiology, 81*(1), 19–25

Clement, G. (2011). *Fundamentals of space medicine* (2nd ed.) New York: Springer (Space Technology Library). ISBN: 978-1-4419-9904-7

Johnston, R. S. (1977). *Biomedical results from Skylab*, NASA SP-377

Johnston, R. S., Dietlein, L. F., & Berry, C. A. (1975). *Biomedical results of Apollo*, NASA SP-368.

Kirsch, K. A., Röcker, L., Gauer, O. H., Krause, R., Leach, C., Wicke, H. J., et al. (1984). Venous pressure in man during weightlessness. *Science, 225*(4658), 218–219.

Nelson, M. (2018). *Pushing our limits. Insights from Biosphere 2.* Tucson: The University of Arizona Press. ISBN:978-0-816-53732-7

Sibonga, J., Matsumoto, T., Jones, J., Shapiro, J., Lang, T., Shackelford, L., et al. (2019). Resistive exercise in astronauts on prolonged spaceflights provides partial protection against spaceflight-induced bone loss. *Bone, 128*, 112037.

Van Ombergen, A., Jillings, S., Jeurissen, B., Tomilovskaya, E., Rühl, R.M., Rumshiskaya, A., et al. (2018). Brain tissue-volume changes in cosmonauts. *The New England Journal of Medicine, 379*(17), 1678–1680.

6

Exploration and Colonisation

Contents

Before starting into this most fascinating field, the reader should remember a fact that puts everything in perspective: No human has ever been beyond Earth orbit for longer than 10 days, and never further than the Moon! To make things worse, the last time a human has left Earth orbit was in 1972. This means, nobody has left Earth's magnetic field during the lifetime of anyone 48 years and older now (in 2020). Mars and near-Earth asteroids have been identified as possible next goals for exploration. This chapter will discuss potential travel destinations and aspects of planning, activities, flight time, survival and resources.

© Springer-Verlag GmbH Germany, part of Springer Nature 2020
B. Ganse, U. Ganse, *The Spacefarer's Handbook*, Springer Praxis Books,
https://doi.org/10.1007/978-3-662-61702-1_6

6.1   Travel Destinations

Aspiring spacefarers may ask themselves which celestial objects are appealing as travel destinations. Due to limited propulsion capabilities, travel outside the solar system seem less realistic (and probably will remain so for a long time). Within the solar system, many bodies, geographical features and anomalies are worth exploring, studying and naming, including mountains and craters on distant worlds (Fig. 6.1). Figure 6.2 gives an overview of the solar system. How long travel would take with current technology is addressed in Fig. 6.3.

Fig. 6.1   Names for planets and geological structures on them, such as mountains or craters, can be suggested on this official website. There are, however, certain restrictions to what names can be given, such as items cannot be named after living persons. http://planetarynames.wr.usgs.gov/

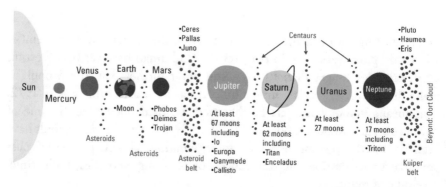

Fig. 6.2   A rough map of the solar system. Distances and sizes are not to scale!

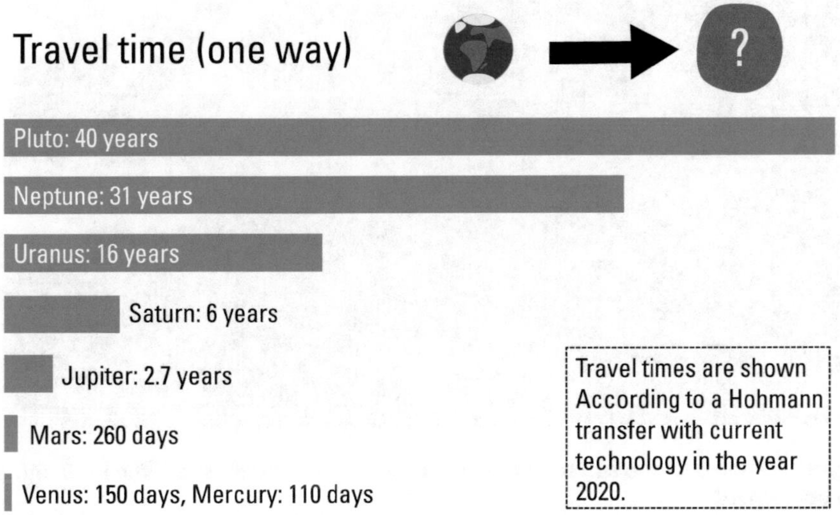

Travel time (one way)

Pluto: 40 years

Neptune: 31 years

Uranus: 16 years

Saturn: 6 years

Jupiter: 2.7 years

Mars: 260 days

Venus: 150 days, Mercury: 110 days

Travel times are shown
According to a Hohmann
transfer with current
technology in the year
2020.

Fig. 6.3 How long does it take to travel to places in the solar system?

Planets are the obvious choice for future exploration missions, especially Earth's nearest neighbour Mars. According to the International Astronomical Union (IAU), an object is a planet, if it

1. orbits the Sun *and*
2. has sufficient mass to develop a spherical shape (hydrostatic equilibrium) *and*
3. has cleaned its orbit from other bodies.

Planets with a rocky surface, such as the Earth, Mars, Mercury or Venus, are called *terrestrial planets*. It is possible to land and move around on them. Due to their proximity to the Sun (and the greenhouse effect on Venus), temperatures on Mercury and Venus are, however, too extreme to allow human missions in the near future. While Mercury has temperatures up to 430 °C in its tenuous atmosphere during the day, Venus is even hotter at 460 °C with an atmospheric pressure 90 times higher than on Earth. In addition, flying to these planets close to the Sun requires a swing-by manoeuvre and more energy than travel in the opposite direction. Space probes have landed on Venus in the scope of the Soviet Venera and Vega programmes (Fig. 6.4). It might be possible to place a balloon-based station into the upper atmospheric layers of Venus, where air pressure and temperature are more moderate. Likewise, the gas giants Jupiter, Saturn, Uranus and Neptune are candidates for the location

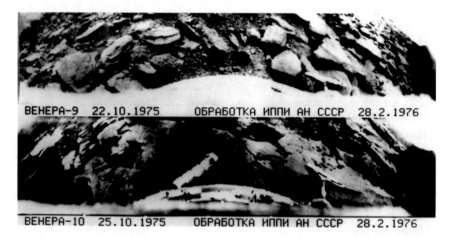

Fig. 6.4 Photos of the surface of Venus, taken during the missions Venera 9 and 10 (Image: NASA)

Fig. 6.5 The landscape of the Kimberley region shows the typical cold desert wastelands on Mars (Image: NASA)

of floating aerostat settlements. These planets lack a clear crust or surface. Instead, the gas density and pressure increase towards the centre, where a solid core is suspected. Study of their interior proves to be difficult, as all probes were destroyed during descent through the gaseous layers. Altogether, when picking a planet for spacetravel, Mars seems to be the most convenient option at the moment. Uncrewed missions have provided plentiful data and photos from the surface (Figs. 6.5 and 6.6). Humans have, however, not been there yet. Mars provides many treasures to explore, especially because it is assumed that it once had an atmosphere and oceans. It is also home to the highest mountain of any

Fig. 6.6 A sand dune on Mars. The photo was taken by NASA's HiRISE camera. Changes in sand dunes over time deliver information on wind and weather (Image: NASA)

planet in the solar system: Olympus Mons. At 21.2 km height, it stands more than twice as tall as Mount Everest on the Earth! (Both mountains are close to the maximum height that their rock foundations can support, but thanks to the weaker surface gravity on Mars, that maximum ends up considerably higher than on Earth.)

It is also possible to land on *dwarf planets* (for example, the asteroids Ceres and Vesta) and Kuiper belt objects beyond the orbit of Neptune like Eris, Haumea and Pluto (Fig. 6.7). Thanks to the *New Horizons* space probe, Pluto is known to have a massive ice coating consisting of frozen nitrogen (98%) and a thin atmosphere likewise containing nitrogen. Only little is known about these bodies at far distance from the Earth. Haumea is another trans-Neptunian dwarf planet (a *plutoid*) with an elliptic shape and a diameter of 2200 km at the equator. The distance between its poles is 1100 km. As there seem to be many such objects, there is clearly need for further exploration here.

The next class of possible destinations is moons. The moons of the giant gas planets should offer stunning views in the sky! They can vary starkly in size and appearance. Standing out among the moons in the solar system are Jupiter's moons Io due to its strong volcanic activity, Europa, an ice planet with a gigantic ocean under the ice crust orbiting Jupiter every 3 days, Ganymede, the largest moon of the solar system, and the only one with a strong magnetic field (while only possessing a thin atmosphere) and Callisto, the second-largest moon of Jupiter with an ice crust and many craters. When considering a flight into Jupiter's vicinity, however, huge and deadly radiation levels in its radiation

Fig. 6.7 Nitrogen ice floes on Pluto pushing against mountains of water ice (Sputnik Planitia and the al-Idrisi mountains). The photo was taken by the *New Horizons* space probe of NASA (Image: NASA)

torus need to be considered. Radiation doses are 5000-times higher than in the Earth's van Allen belts (Fig. 2.3), which means that even a quick crossing of one of these belts would be lethal for humans. Only Callisto can be reached without crossing these dangerous areas. This example, again, underlines the importance to develop radiation protection technology for spaceships.

Saturn has a significantly weaker radiation environment since its rings capture trapped high-energy particles. This gives the Saturnian moons a more survivable environment and makes them more interesting destinations despite their relatively larger distance from Earth. Saturn's moon *Titan* is the object in the solar system with the most similarities to Earth. It has a thick atmosphere containing nitrogen gas, and water ice and hydrocarbons have been found to make up its surface layers by the Cassini–Huygens mission. Experts discuss the possibility of life on Titan.

Fig. 6.8 The Cassini spacecraft in 2011 took this photo showing from left to right the Saturn moons Janus, Pandora, Enceladus, Mimas and Rhea. A part of the A and F rings is visible. The rings and all moons are orbiting in the same plane (Image: NASA)

The rings of Saturn are almost one million kilometres in diameter, since they surround the entire planet, but only 100 m thick. They consist of so-called dust with an average particle size of 5 m and a distance of 10–100 m between objects, constantly in motion. Their material is almost entirely (99.9%) made up of water ice, with only a small fraction of rock inclusions. As particles move relative to each other with speeds of several centimetres per second, they repeatedly collide, thereby undergoing a constant mechanical grinding process. The origin of the ring matter is most likely a former icy moon of Saturn that broke apart. Figure 6.8 shows several of Saturn's moons and the rings.

Further exciting and possibly economically interesting destinations are *asteroids*. These small rock bodies have a weak gravitational pull and their shape does mostly not develop into a sphere. Millions of them exist in the solar system. The largest number can be found in the asteroid belt between the planets Mars and Jupiter. Distances between them are on the order of millions of kilometres and not, like typically shown in science-fiction movies, close together. Risk of collision is very low! In case a collision occurs despite its low likelihood, fragments might later hit other planets as meteoroids.

Asteroids are classified by their location in the solar system, their structure and composition. The three major spectroscopic classes, according the SMASS classification, are *C-group* (Carbonaceous asteroids, rich in carbon, 75% of asteroids), *S-group* (Silicaceous or stony asteroids, 17%) and *X-group* (Metallic and other asteroids). Most asteroids are solid rocky objects, others only consist of ice and dust or stones pulled together by their gravity (so-called rubble piles). Ceres is the largest object of the asteroid belt. Its surface is made of silicates and carbon (making it a C-group object). Several white spots on its surface have been identified as salt deposits from evaporated cryovolcanic water.

Grouping by orbital location gives further structure to the asteroid zoo: Some asteroids collect in the Lagrangian points of planets (compare Sect. 3.5). These are called *Trojans*. Examples of Trojans are *Achilles* (Jupiter) and *Eureka* (Mars). Some asteroids have moons of significant size. Examples include asteroid Moshup, a near-Earth object of the Aten-group that orbits the Sun every 6.18 months, orbited by a companion named Squannit. The pair of Didymos and Didymoon, target of the AIDA asteroid deflection mission is likewise noteworthy. The asteroids in the main belt between Mars and Jupiter are classified in three groups: an inner, middle and outer belt. Further grouping is based on particular orbital properties, such as the Cybele group or the Pallas family. Near-Earth asteroids are those with close encounters to the Earth orbit. Due to their proximity, they can be reached in relatively short travel times. The classification considers four types of near-Earth asteroids: the Atira group (completely inside of Earth's orbit), the Aten and Apollo group (crossing Earth's orbit) and Amor group (completely outside of Earth's orbit). In addition, co-orbital asteroids share a similar and stable orbit with Earth but never come close.

Comets are icy and dusty bodies from the outer reaches of the solar system that travel on highly elliptical orbits. Landing on a comet and travelling with it allows a multitude of solar system environments to be studied, as both local physical conditions and observation opportunities of other nearby objects vary over time. Temperature changes lead to evaporation, outgassing or melting as comets get closer to the sun. The evaporated gasses are pushed away by the solar wind flow and radiation pressure, leading to the formation of a coma and tail. The tail hence always points away from the sun, as it is shaped by the solar wind flow rather than the motion of the comet. The coupled ESA space probes *Rosetta* and *Philae* were the first in human history to orbit a comet, with Philae continuing to land on *67P/Churyumov–Gerasimenko*. Fitted with sensors and cameras, it delivered fascinating images and data from the comet's surface on its trip around the Sun (Figs. 6.9 and 6.10). Periodic comets have an orbital period of 200 years or less, while most aperiodic comets never return to the

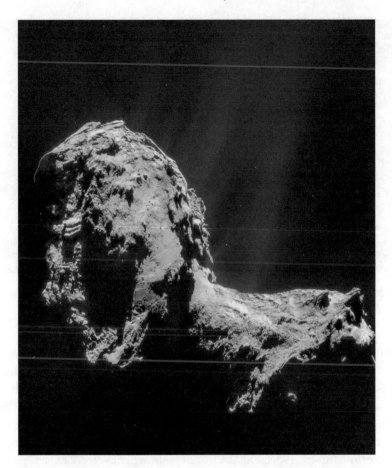

Fig. 6.9 Comet 67P/Churyumov–Gerasimenko. The photo was taken by the European Space Agency's *Rosetta* space probe. Outgassing can be seen in the comet's atmosphere (Image: NASA)

solar system at all and continue on a hyperbolic trajectory. It is assumed that the objects in the Oort cloud are of the same composition as comets (Fig. 6.2).

Even though current technology does not allow human travel outside the solar system, the travel guide of potential destinations does not end here. When leaving the solar system, a density jump is encountered where the solar wind accumulates, called the *termination shock*. Even though particles density is very low, the termination shock behaves just like the bow wave in front of a boat in water. The spaceship itself, just like any other objects in the solar system, likewise causes a bow wave of this type, as it travels through the solar wind. At the termination shock, the flow direction of the surrounding

Fig. 6.10 Photo of the surface of comet 67P/Churyumov–Gerasimenko taken from the lander *Philae* (Image: NASA)

medium changes abruptly, and the solar wind stops flowing away from the Sun. It starts mixing up with the interstellar medium. Both media consist of charged particles but differ in density, temperature and velocity. Another few million kilometres further out, another border is reached, where the radiation intensity suddenly increases. At this outer edge of the solar system, the Sun is no more the dominating radiation source. Instead, the radiation coming from other cosmic sources, especially the pulsar Geminga now reaches the spacecraft without any deflection from solar wind influence. The *Voyager* space probes were launched in the 1970s and are still operational today. They are the only source of information about these far-flung regions and send the most valuable data while they continue their journey away from the Earth.

The solar system is part of the Milky Way galaxy, specifically embedded in a low density area called the Local Bubble. Inside it, the solar system has crossed a cloud of interstellar matter with a diameter of 30 light years, called the Local Interstellar Cloud (LIC), or Local Fluff, in the last 100,000 years. The density of the interstellar matter in the Local Fluff, however, is very low: one atom per four cubic centimetres. Our neighbouring visible stars are Alpha Centauri (only 4 lightyears away), Sirius A (8.7 lightyears) and Epsilon Eridiani (10.4 lightyears). New stars are continuously formed in regions of high-density interstellar matter, the closest one being the Rho Ophiuchi cloud at 140 light years distance.

One of the hottest topics in current astronomy is exoplanets, meaning planets orbiting around other stars than the sun. In recent years, thousands of extrasolar planets have been identified, always with the implicit hope to find any evidence of extraterrestrial life forms. The first planets outside the Solar System were found indirectly in the 1990s, through periodic dimming of their host stars as they were eclipsed by the planets travelling in front of them. The first directly imaged exoplanet was Fomalhaut b (also called Dagon), which is about the size of Jupiter and located at a distance of 25 lightyears. Exoplanets have also been identified in the *habitable zone* of other solar systems, which means they are in a distance to the star that allows for the existence of liquid water. This is often considered necessary for life as we know it to evolve. In our direct cosmic neighbourhood, orbiting the star Proxima Centauri, an Earth-like planet was found in the year 2016, called *Proxima b*. It is only 4.2 lightyears away!

Further planets in the habitable zones of their systems are, for example, Kepler 22b and Gliese 581c. Kepler 22b is approximately 600 lightyears away and much bigger than Earth, making it a so-called super Earth. This designation does not imply judgement of living or climate conditions there, it just addresses the fact that the planet is larger than Earth, but smaller than Neptune. Even larger Earth-like planets are called *mega-Earths*. Gliese 581c is only 20 lightyears away from Earth and seems to be a super Earth. Planets orbiting a star are named in order of their discovery, which can be arbitrary, and start with the letter b. It is not possible to infer the orbital ordering from the planets names themselves! Among the known types of exoplanets, there are also some only consisting of ice and dirt, similar to comets in the solar system. It is possible to imagine such a planet when considering how water forms a sphere in weightlessness (Fig. 6.11).

In between stars, objects the size of a planet are assumed to exist, that do not orbit a larger body, called *planetary-mass object* (PMO), or *planemo*. When a planet gets thrown out of its orbit (either by the supernova explosion of its host star or a gravitational interaction with another body), it can become a *rogue planet* with an orbit around the galactic centre of a galaxy. It is also called interstellar or nomad planet. As these planets are not close to a star that emits light, they are dark and cold, and therefore hard to spot.

Fig. 6.11   NASA astronaut Chris Cassidy with free-floating water between him and the camera (Image: NASA)

6.2   Exploration

Before embarking on a mission to a planet, moon or asteroid, it is important to gather as much information as possible to select an appropriate landing site. Once having arrived, the next step is to explore the surrounding area. Depending on the terrain and strength of gravity, it might be helpful to have a vehicle or an uncrewed exploration option, such as a robot, mobile surface probe or drone available. Surface properties may vary, and therefore it is better to have several alternatives at hand. Collection, storage and analysis of samples should be possible. To be able to search for life, sterile sample handling is required and contamination of the site with terrestrial organisms should be avoided as much as possible.

Already during the last three missions of the Apollo programme, Apollo 15, 16 and 17 (Figs. 5.3 and 6.12) lightweight rovers were used on the Moon. These battery-powered *Moon buggies* only weighed 210 kg each and increased the exploration range enormously, travelling at a speed of up to 13 km/h. As gravity on the Moon is only a sixth as strong as the Earth's gravity, the tires did not need to be solid, but it was enough to build them of wire mesh (Fig. 6.13). As only one vehicle was available on each mission, and it was necessary to be

Fig. 6.12 Astronaut Eugene A. Cernan tests the *Lunar Roving Vehicle* during the Apollo-17 mission in December 1972 (Image: NASA)

able to walk back in case of failure, the radius for exploration was limited to 6 km from the lander. Like in any other stage of a spaceflight, redundancy and failure safety need to be considered when planning exploration. If possible, two vehicles should be taken to enable for rescue missions. Depending on the location, a suitable source for generation of electric energy (be it solar, geothermal or nuclear) needs to be considered. The Apollo rovers were left behind on the Moon and are still there today.

In the Soviet space programme, the *Lunokhod programme* (Russian for moonwalker) was a great success. In 1970 and 1973, Lunokhod 1 and 2, respectively, landed on the Moon. These remote-controlled robots were fitted with cameras and instruments, including X-ray spectrometers, radiation detectors

Fig. 6.13 Tire of the *Lunar Roving Vehicle* in lightweight construction. It was stored on the side of the lunar lander and had to be assembled before use (Image: B. Ganse)

and telescopes. In recent decades on Mars, remote-controlled rovers were used more frequently. In the 1970s, the Soviet Union tried to bring two rovers to the red planet with the missions *Mars 2* and *Mars 3*. However, both missions failed. The rover *Sojourner* of NASA's Mars Pathfinder mission in 1997 was the first rover to be successfully used on the Mars surface. Only a few years later, in 2004, the identical NASA rovers *Spirit* and *Opportunity* followed, and in 2012 the *Mars Science Laboratory Curiosity* started to explore Mars (Fig. 6.14). These rovers delivered valuable research data by analysing air, soil and rock composition and gave humankind a great impression of what the Mars surface looks like. The rover Opportunity drove more than 40 km in total, despite being stuck in sand dunes several times (Fig. 6.6). The Japanese Hayabusa2

Fig. 6.14 The Mars rover *Curiosity* in the Martian landscape. Quiz: Who took this photo? The rover used its robotic arm to take several photos of itself. These were later stitched together into a complete picture (Image: NASA)

space probe deployed four rovers on the surface of the asteroid Ryugu, which moved in the low surface gravity by hopping. The most distant object that has so far been explored by a space probe that landed on it is Saturn's moon Titan. The Huygens probe of the *Cassini–Huygens mission* landed there in 2005.

Crewed rover vehicles need to have the following features: they are required to be lightweight, suitable and agile in the target terrain, capable of crew and

Fig. 6.15 NASA's SEV (*Space Exploration Vehicle*) (Image: NASA)

instrument transport and carry a sufficient amount of fuel, or use an energy source that can be provided on-site. It is in principle possible to have the crew either wear a spacesuit while driving or sit in a pressurised chamber fitted with a life support system. During the Apollo missions, EVA suits were worn while sitting on the rover, as other systems were not applicable due to payload restrictions.

NASA plans to use vehicles with a pressure chamber in the future that can be left only wearing EVA suits. The *Space Exploration Vehicle Concept* (SEV) intends to use the same chamber on a rover *Surface Concept* (Fig. 6.15) and in spacecraft, where boosters and control thrusters or maybe even robotic arms are attached for the *In-Space Concept*. The fundamental concept to use a modular vehicle in different situations to save payload makes a lot of sense. Another aspect of it is to reduce development costs by utilising modular systems.

6.3   Volcanism, Glaciers and Permafrost

When planning colonies and exploration missions, the landing sites' local geography needs to be taken into account. Multiple bodies in the solar system possess volcanic activity, such as Jupiter's moon Io (Fig. 6.16). It is fuelled by

Fig. 6.16 Jupiter's moon Io with volcanic activity photographed by the *Galileo* space probe. The tidal forces on Io are 6000 times as strong as on Earth due to the great mass of Jupiter, causing volcanic processes (Image: NASA)

the planets deformation from tidal forces. Volcanism on Earth has a totally different source: radioactive decay heat in the core of the planet. Volcanic activity on the moon Io includes volcanic eruptions, lava streams, columns of smoke reaching up to several hundred kilometres of altitude and pyroclastic flows. Eruption temperatures of up to 1300 °C have been measured. Fast-moving pyroclastic flows are not only dangerous for spacefarers, but also the main reason for deaths in volcanic outbursts on Earth.

On several bodies of the solar system, volcanism occurs with a cool twist: on icy moons of the giant gas planets the volcanic activity takes place at the melting point of fluids (such as water or methane) instead of rock. This is called *cryovolcanism*. Examples are Neptune's moon Triton and Saturn's moon Enceladus.

Ice can be found on many planets and moons (Fig. 6.7). On Earth, water ice is most common, but many other types of ice are present throughout the solar system, such as methane ice, frozen CO_2 and frozen nitrogen. Ice may float on an ocean, be bound in soil or rest directly on solid ground. On Earth, *glaciers* are usually harmless, but a few things need to be considered: water ice glaciers that have not moved for a longer period of time may suddenly advance quickly at up to 50 m per day, developing very high forces. This is the case when meltwater channels under the glacier are blocked and a large amount of basal water pressure accumulates. In a sudden surge, the tension gets released and the ice moves forward. The hydrostatic pressure supporting this phenomenon depends on the gravity of the object—the stronger the gravity, the higher the pressure. Glaciers of methane ice or other ice types behave differently than glaciers on Earth due to different density, friction, buoyancy and compressibility, if they exist at all. In the proximity of glaciers, landscape changes can occur unexpectedly. *Moraines* are deposits of sediments left behind by glaciers, containing a large spectrum of grain sizes. Glaciers can also leave enclosures of ice that melt away slowly, and eventually collapse when the surface ice itself is long gone, forming a kettle. Ground around them can remain unstable for some time. On bodies with weak gravity, this problem might even be more pronounced than on Earth. For these reasons, colonies should preferably not be built close to glaciers.

On colder planets and moons with seasons, *permafrost* can be a problem. When the ground is frozen year-round and only a superficial layer of mud and water defrosts, spaceships may get bogged down in mud. Fluid collects on top of the still frozen layers of soil, as the frozen ground cannot absorb it (thermokarst). In this case, also exploration vehicles need to be appropriately designed to not be limited in their motion. Sunk-in objects may freeze in the soil during the colder season and it might be impossible to move them. Landing-spots need to be selected accordingly, for example, at an elevation or a small hill.

6.4   Human Colonies

Mankind has been dreaming of building colonies on other celestial bodies for centuries, especially on Moon and Mars. *Colonisation* of foreign worlds has the fascination of creating a human outpost for research, commercial and touristic purposes. In times of increasing environmental destruction and climate change on Earth, the idea of an outpost that guarantees survival of the human species in case of a serious catastrophe on Earth is tempting. This would, however, only be possible, if the outpost was entirely independent. When European settlers colonised the American continent, they found it already to be inhabited with good living conditions, food, water and animals. As no such things can be found outside the Earth yet, it will be much harder to reach autonomy. The Moons and other planets of the solar system neither have an atmosphere breathable for humans nor plants or ecosystems available. For this reason, concepts to produce food and regenerate air need to be developed. As most planets and moons do not have an atmosphere or magnetic field, another important topic is protection from radiation. On the Moon, for example, cosmic radiation directly reaches the surface and hits the atoms of the rock, creating secondary particles. This results in the curious situation that the radiation dose has its maximum in 1 m of depth under the surface. When planning a colony that protects its inhabitants against radiation, a thicker surface of rock as a cover, water or magnetic shielding need to be considered. In addition, temperatures comfortable to humans would be desirable, as extreme temperatures require more protection and a higher energy consumption for regulation (see Sect. 2.4).

To be able to survive breathing an *atmosphere*, humans need a suitable air pressure, oxygen content and oxygen partial pressure (warning: danger of fire when too high). A low carbon dioxide level and no toxic substances should be in the air. Cancer or lung diseases can develop over time when exposed to toxic air. Lung fibrosis can develop when inhaling dust particles for long durations. Life support systems were already described in Sect. 2.5 and medical phenomena discussed in Sect. 5.5. The chapter on plants and terraforming will present further options to create favourable atmospheric conditions (Sect. 6.5).

The average air pressure on Earth is 1000 hPa at sea level. For humans, however, depending on the air composition, much lower pressures suffice. An adaptation time of up to several weeks might be needed to be able to cope with the conditions. The other atmospheres in the solar system differ largely from the one on Earth. Humans are only capable of breathing the Earth's air—all others would be deadly within a short time. Oxygen is present in the Earth's atmosphere (21%), on Mercury (42%, but vanishingly low air pressure) and on Jupiter's Moon Europa (100%, also very low pressure). Humans need a pressure suit on Mercury, Mars and Europa. A similar air pressure level to that on Earth can be found on Titan (approximately 1500 hPa). However, the atmosphere consists of 98.4% nitrogen and 1.6% methane and argon. Due to the absence of oxygen, humans cannot survive breathing this atmosphere.

The thin atmosphere of Mars today consists of 95% carbon dioxide (Fig. 6.17). However, scientists found evidence that liquid water once existed on Mars, and that it probably had a denser gaseous shell with a different composition. Some researchers even believe the atmosphere has contained oxygen. It is currently thought that three mechanisms might have contributed to the atmospheric loss over a very long period of time. Firstly, the solar wind

Fig. 6.17 The atmosphere of Mars. Photo taken by the Viking 1 space probe (Image: NASA)

might have slowly carried the molecules away. Since Mars does not possess a magnetic field, ionised particles are not trapped but escape easily. Secondly, a larger collision (such as by a meteoroid) could have been the cause, or thirdly, the atmosphere may have slowly disappeared because the gravity of Mars is much lower than on Earth. Nowadays, its surface is completely formed by dry rock and dust, with a clearly red colour due to a high content in iron oxide. In the absence of water, Martian dust grains are particularly nasty, with sharp edges and toxic chemicals such as perchlorates on their surfaces. They could wear out spacesuits rather quickly and cause lung problems when inhaled. Water ice was found to be up to 3 km thick in the *polar ice caps*, partially covered by carbon dioxide ice, called *dry ice*. On the Martian south pole, the thin cover of frozen CO_2 is 8 m thick. In 2018, a subglacial lake was found

1.5 km below the surface and with a diameter of 20 km, that needs to be explored further to learn more about Mars, its history and potential life forms.

When considering to stay and move around on another celestial body, the strength of its gravitational attraction has a big influence on movement. Gravity on the surface of Earth is defined as 1 G ($= 9.81$ m/s^2). The Apollo astronauts on the Moon experienced what it is like to move in an environment with only one sixth G. Their preferred way to move, instead of walking was to jump. Jumping was apparently much easier and more efficient than walking. For this reason, it needs to be assumed that humans will adapt their motion to the level of gravity, and maybe even crawl on all fours in higher gravity conditions. This aspect should be considered in spacesuit and equipment design. As Mars has about a third of the Earth's gravity, it might also be more convenient to jump instead of walk. In case of very high gravity levels, the human body will probably need to adapt for a longer period of days to weeks. It will not be possible to walk normally straight away, especially, as the blood will collect in the legs and cause orthostatic problems. In addition, the muscles will have to work very hard to maintain posture. In parabolic flights and human centrifuges, lower and higher gravity levels can be simulated. In the centrifuge, healthy humans can cope with 9 G only for a very short period of time. The authors recommend from their own experiences in human centrifuges to not land on a body with more than 3 G. Each individual, however, has their own susceptibility level to G-forces, and it might be advisable to use daily centrifugation to exercise before landing on a body with higher G-levels. It is also likely that humans would adapt to life in a high-G-environment with stronger muscles, bones, hearts and adaptations of the blood vessels. This topic has, however, so far not been in the focus of research.

Architects, artists and designers have created a multitude of design concepts for lunar and Martian bases and colonies. Overall, the main two types of concepts are over- and underground bases. All of them need to include the same basic systems as spaceships, including life support, waste management, water recycling, storage and much more. Living, working and private spaces are as important as more specific systems accommodating for mission-specific tools and vehicles. From isolation studies, such as the Biosphere-2-experiments (Sect. 5.9), disadvantages of glass cupolas and concrete structures are known. Current concepts rely more on modular structures, possibly under a thick layer of dust and rock designs based on 3D printing are under development. Plants can be grown in facilities sporting artificial LED light sources, nutrient delivery and a suitable substrate, in a vertical farming setup. Inflatable modules may be an option, as they are more lightweight and may be delivered more easily than heavier solid structures (Fig. 6.18).

Fig. 6.18 Example of an inflatable spacestation module built by Bigelow Aerospace. The photo was taken in building 9 of the Johnson Space Center (Image: B. Ganse)

6.5   Plants and Terraforming

Food production will be a central aspect for the success of human colonies. Nutrition needs to be diverse and provide all necessary nutrients, including enough protein, vitamins and healthy fats. To meet psychological demands, as mentioned in Sect. 4.3, food also needs to be tasty. An unbalanced diet may lead to deficiency symptoms. As an example, the Crusaders in the middle ages, the sailors during the Age of Discovery in the early modern period, and the arctic explorers in the nineteenth and twentieth century suffered from *scurvy*. It occurs after 2–4 months without vitamin C intake and typical symptoms include bleeding gums, skin infections, wounds, fatigue, joint pain, fever,

Fig. 6.19 A blooming zinnia in the Veg-01 experiment on the ISS, testing a plant growth system using plant growth pillows (Image: NASA)

depression, cardiac weakness and diarrhoea. The best way to avoid vitamin deficiencies is to eat fresh vegetables and fruit, and add vitamin tablets if unable to provide enough of them. For these reasons, plant cultivation facilities are desirable for colonies, and would also increase crew comfort and health on long term spaceflight. Experiments of plant cultivation in weightlessness and in closed habitats have been conducted for a long time. Already in 1982, rockcress was grown on board the Soviet spacestation *Salyut 7*. Recently, *youth-and-age plants* (Zinnias), salad and other plants were grown on the International Space Station (Figs. 6.19 and 6.20). Plants can be grown without soil in the so-called *hydroponics systems*, a type of hydroculture. Nutrient solutions are added to water. The plant roots will either grow directly into the water or be supported by stones or gravel. In self-sustaining aquaponics systems, fish and bacteria help to keep up the nutrient cycle for the plants. Light is essential for plant growth and needs to be intense enough for sufficient results.

The next question is which plants to grow in a spaceship or in a human outpost. The scientific discipline dealing with this topic is called *astrobotany*. Plants should be suitable and high-yielding. Experiments on the ISS have shown that plant growth in weightlessness is feasible. However, this might not be true for all plants to the same degree, which is why more research is

Fig. 6.20 Astronaut Steve Swanson cultivating salad on the ISS (Image: NASA)

needed. When searching for suitable options to grow food and complement the crew's menu, a variety of fast-growing fruit and vegetables could be an option. Several groups of scientists are currently working on this problem. In a project funded by the European Union, food growth is tested in a greenhouse in Antarctica (project *EDEN ISS*). It uses a partly aeroponic closed water cycle system without soil, optimised LED light and enriched carbon dioxide levels. With more CO_2 in the air, photosynthesis is more effective and plants show increased growth rates. Cucumber, salad and tomatoes produce good yields. Among the available systems, the German Aerospace Center has developed the C.R.O.P. (Combined Regenerative Organic food Production) concept. It seeks to use organic waste for plant growth and thereby food production.

On Earth, a huge variety of seeds exists. To preserve this variety, seed banks exist around the world. In 2008, a seed bank was opened on Svalbard in the Arctic, called the *Svalbard Global Seed Vault*. It is capable of storing 4.5 million seed specimen from all over the world in an underground bunker at −18 °C. In case of serious species-endangering catastrophes, these seeds could allow re-cultivation (Fig. 6.21). Such a seed vault could in theory also be installed on the Moon to contain the Earth's seed variety. It is important to keep seeds cool and protected them from radiation.

Fig. 6.21 Entrance to the *Svalbard Global Seed Vault*, the largest seed bank in the world located near Longyearbyen on Svalbard in the Arctic. Plant seeds from all continents of the Earth are systematically collected here. In case of a serious catastrophe, the seed vault wants to help against the loss of biodiversity (Image: B. Ganse)

Another topic of great public interest is *Terraforming*. The idea to transform an uninhabitable planet into an Earth-like place has been considered under this term since 1942. The plan is to alter atmospheric and chemical processes in a way that results in a breathable atmosphere and plant growth. Terraforming is currently far from being feasible. In theory, depending on the initial situation of the planet or moon, the application of several entirely separate procedures might be necessary to eventually reach conditions that allow life to flourish. Potential strategies include creation of an atmosphere, introduction of plants, inhibition or enhancement of the greenhouse effect, conversion of CO_2 and generation of O_2 (for example, by the *Sabatier reaction and electrolysis*, see Sect. 2.5), or altering the orbit around a star. The Martian regolith could release oxygen in the presence of CO_2 and water. In science fiction, terraforming is a recurring feature, that is, however, currently not realistic, due to the large amounts of energy needed.

A popular, but currently also unrealistic concept is the *generation ship*. The idea is to let a big group of people travel a very long distance, while the trip takes longer than their lifetime. Their children or grandchildren will eventually arrive at the destination and hopefully be able to accomplish their mission goals. The spaceship is required to be large enough to allow for a comfortable life on board. As children need to be born and raised on board, many facilities and enough room are necessary. Food and medications, as well as components for repair need to be produced on board. In addition, it should be possible to receive updates and build new technological developments coming from Earth. Social life and culture require a lot of attention, too. A gigantic database would supply the crew with all available knowledge. Experts of all relevant fields need to pass on their knowledge and skills to the next generation. A later generation might decide to return to Earth, and this option should be available to them at any time. Genetic diversity needs to be provided to avoid medical issues due to inbreeding. Methods are required to strengthen the immune systems of those born on the spaceship, maybe by immunisations. Experience from isolated groups of people on Earth shows that culture quickly develops in its own direction. For this reason, re-integration after return might be very difficult not only for medical, but also for social reasons.

6.6   Resources and Mining

On Earth, mining and resource extraction takes place even in remote areas and despite all obstacles. The logical conclusion is that mining will also be performed in space. Certainly, not any type of resource is worth the effort of flying to an asteroid, moon or a different planet; but some rare and highly valuable elements, such as iridium, could justify the expense. Apart from the decision whether the resource is financially interesting enough, also the technical abilities to mine it need to be available. Transport capacity and drilling or mining tools, as well as rocket technology will need to be significantly advanced. In principle, two types of mining exist on Earth: surface mining and underground mining. For liquid and gaseous resources, bore holes and pumps will do the job, as does extraction of liquids in substrates, such as tar sands. In space, direct collection of resources on the surface would be preferable, or entire small celestial bodies could be brought into an Earth orbit.

So far, mining has not commercially taken place outside the Earth. On the Moon, the Apollo astronauts drilled rock core samples for scientific analysis. The same was done by uncrewed probes on other bodies such as Mars, where the sample analysis was performed on-site. On the Moon, a rock layer called KREEP (K for potassium, REE for *Rare Earth Elements* and P for phosphorus) was discovered, that is usually hidden beneath the crust, but brought to the surface by impact events in the form of breccia. Since it is rich in rare earth elements, KREEP is an interesting example for a mineable resource on a celestial body. Lunar basalt and troctolite were also found and brought back to Earth by the Apollo missions. Both rock types are very old, aged around 3.2–4.4 billion years. Likely, more interesting resources are available on the other planets, asteroids and moons in the solar system. Asteroids seem to provide a great amount of minerals on the surface. Besides rare earth elements, also gold, silver and platinum group metals were found from asteroid samples. Asteroids have a great advantage for mining activities: gravity is so low that only little fuel is needed for departure. The rings of Saturn might be another great source for mining activities, as, apart from a variety of rock types, also water ice is abundantly available there. Concepts for mining missions currently primarily involve robotic prospectors, while no crewed mining is seriously considered.

Even though mining missions would focus on business and earning money, it is important to also allow for research and involve science institutions. Especially astropaleontology and the search for signs of extraterrestrial life are important to consider. Once space mining becomes commonplace, regulations might be required to preserve the wonders of the solar system for touristic and heritage reasons.

6.7 Space Law

From a legal perspective, it is quite difficult to define where space begins. While all countries agree that their legislative authority ends at a certain height above the surface, there is no international agreement as to the precise altitude, where the boundary between a countries' airspace and space should be set. Altitudes of 80 or 100 km are common in some countries' law texts and so are definitions that space begins at a height, in which aerodynamic flight becomes impossible. Above that level, the legal framework governing space is somewhat akin to the rules of the high seas: Spacecraft launched by a country follow the laws of that country, and the commanding officer of the craft enacts laws of that jurisdiction, much like the captain of a ship sailing in the open ocean. This implies, for example, that abandoned spacecraft can be treated as flotsam and jetsam and may be commandeered if found.

Five international legal frameworks have been defined by the United Nations to regulate these matters further: The "Treaty on Principles Governing the Activities of States in the Exploration and Use of Outer Space, including the Moon and Other Celestial Bodies" (Outer Space Treaty) was signed in 1967, requiring states to use space only in the interest of all mankind and for peaceful purposes.

The "Agreement Governing the Activities of States on the Moon and Other Celestial Bodies", also known as the Moon Treaty or Moon Agreement, created in the year 1979, specifies that military use of the Moon and other celestial objects is prohibited. Additionally, it prohibits sovereign countries from claiming or distributing areas or private property on the Moon. Instead, the use of extraterrestrial resources and land usage rights are regulated by the United Nations, with the goal to ensure the "common heritage of mankind". All big spacefaring nations have so far refused to sign or ratify this contract, so it has little of actual relevance. Further space treaties include the Registration Convention, demanding that every launch of a spacecraft needs to be registered with the United Nations (overall, 88% of launched objects have followed this registration demand).

In case of calamities or accidents in spaceflight, the "Rescue Agreement" of 1968 stipulates that all countries need to assist and rescue spacefarers in distress and return them to their home country, in case they land outside of it.

The Space Liability Convention establishes rules beyond the framework of the Outer Space Treaty, for compensation of damages caused by spacecraft accidents. This is a treaty between state actors and can only be invoked by a state suing another state, while the liabilities of organisations and individuals have to be handled by each country separately. This convention has been invoked once in 1976, when the Russian nuclear powered Kosmos 954 satellite crashed into Canada. Russia was charged for the resulting cleanup operation.

The most important take home message for aspiring spacefarers is that the laws of space are very similar to the ones of the high seas. One important difference is that no ground can legally be claimed simply by ramming a flag pole into the soil. So if somebody offers real estate on the Moon: This is simply not legitimate and that person is a fraud!

Quite a number of legal questions in space are still completely undefined, as no precedent has occurred for them: Which laws and customs apply, if something happens when spacecraft from different countries are docked together? Who is liable if spacecraft collide or damage each other? Under which condition can a colony on another planet declare independence from Earth?

6.8   Astrobiology and Extraterrestrial Life

First: despite decades of active searching, no proof of extraterrestrial life has been found! There has neither been contact with aliens nor were any living beings found, despite strong theoretical arguments that favour their existence. This includes organisms on a cellular level, viruses, bacteria, fungi, plants and any other form of life. No biosignatures in the form of clearly alien radio or light signals have been received either. The idea of *cosmic pluralism*, meaning the existence of many inhabited worlds like Earth, however, dates back to the ancient Greek debates, medieval Muslim scholars and Hindu cosmology.

Excursion

The existence of extraterrestrial life has been an object of speculation for many decades and led to divers theories. While Copernicus postulated *heliocentrism* in the sixteenth century, the Italian philosopher Giordano Bruno assumed a never-ending universe, and that stars are suns, orbited by planets (Zekl and Copernicus 2006). The catholic church, seeing its authority threatened, sentenced him to death for heresy. In the seventeenth century, the Dutch astronomer Christiaan Huygens suggested life on other planets and published his *Vernünftige Muthmaßungen* (reasonable speculations) (Huygens 1698). Immanuel Kant, in his 1755 text *On the inhabitants of the stars* published the theory that the cognitive ability of life forms increases with the distance to the Sun (Kant 1755).

All currently known life forms have bodies based on hydrocarbon compounds and water. Carbon is particularly suitable for life due to possibility for four covalent bonds that enable for three-dimensional structures. In theory, other elements with tetrahedral bonds such as silicon, could serve the same purpose, however, no life forms based on anything else then carbon have so far been found. Life on Earth consists, apart from carbon and water, mainly of nitrogen, oxygen, phosphorus and sulphur. The *Gaia hypothesis* proposes that each planet with life has an atmosphere that stays in equilibrium with the planet's inhabitants. This would allow life to be found from far away, by detecting its traces in the atmosphere. This hypothesis is, however, highly controversial, as all sorts of scenarios are conceivable, where the situation is different. This is in particular true, as animals have been found that do not absorb free oxygen from their living environments. Life is also possible and has been found in habitats completely isolated from the atmosphere, such as in rock layers or under ice shields.

In the solar system, among the most promising places for possible extraterrestrial life is Jupiter's moon Europa. A huge ocean seems to be located under the ice surface, heated by geothermal activity from tidal forcing. Cracks and holes have been observed in the ice crust, allowing for an exchange of matter and energy between the surface and the ocean. For these reasons, experts consider life on Europa possible. Further candidates for life in the solar system include the Saturn moons Titan and Enceladus. On Titan, the Cassini–Huygens mission has found methane lakes, filled with liquid methane and ethane (they might melt plastic, so care should be taken when landing, Fig. 6.22). They were the first lakes with liquid content found outside the Earth. Water vapour was discovered on Enceladus, ejected by geysers. Similar to Europa, Enceladus also appears to have liquid water collections or oceans under the ice crust, which are evidenced by these geysers. It remains highly

Fig. 6.22 Methane lakes on the surface of Saturn's moon Titan. Cassini radar image taken in 2006 (Image: NASA)

exciting what further research on these three candidates will deliver over the next years to come.

> **Anecdote**
>
> When geysers were discovered on the Saturnian moon Enceladus by the Cassini space probe, scientists at a NASA press conference proudly proclaimed to have discovered water, CO_2 and traces of hydrocarbons in their ejecta. They jokingly concluded that they had found a beer volcano.

When searching for signs of life on exoplanets outside the solar system, the *habitable zone* plays an important role. Location and extent of the habitable zone depend on the temperature and luminosity of the star. In addition, the surface properties and albedo (amount of reflection of the planet) are relevant, as they determine how much irradiance actually reaches the planet's surface. Throughout the life of a star, the habitable zone moves outwards as its luminosity increases. If life requires a long time to evolve, the planet needs to be continuously located in the habitable zone to make this possible. A greenhouse effect can affect the temperature on the surface. In the presence of greenhouse gases, short-wave radiation can easily penetrate the atmosphere, while long-wave infrared radiation light from the surface is prevented from escaping through the atmosphere, thus trapping heat. In the solar system, according to the classical definition, Earth and Mars are located in the habitable zone. By now, however, scientists have come to the conclusion that life may also exist in other areas outside the habitable zone. These include the aforementioned moons, where heat is produced by friction caused by tidal forces. In addition, radioactivity can provide temperatures allowing for life.

Currently, four types of habitats are distinguished with decreasing likelihood of supporting life in familiar forms:

1. Earth-like planets in the habitable zone orbiting a star.
2. Planets in or on the edges of the habitable zone, developing differently than Earth.
3. Moons or planets with an ocean under the surface, heated by tidal forces or radioactivity, allowing for liquid water.
4. Ocean/water habitats under ice, where high pressure causes water to be liquid despite low temperatures.

In addition, life could exist on exoplanets that are only sometimes in the habitable zone, as their orbit is elliptic. Possibly, life forms could hibernate or transform to resistant survival states, while the climate is less comfortable.

Stromatolites or *stromatolith* have existed on Earth already 400,000 years after its birth. These life forms have already populated Earth 3.5–4 billion years ago. In addition to fossil records, living colonies still exist in a few isolated environments (usually with high salinity). Stromatolites are among the oldest known life forms. They are colonies of cyanobacteria, creating characteristic rock formations (Fig. 6.23). A central point of scientific speculation is, whether these life forms came into existence on Earth by chemical processes in this very short period of time by themselves, or if they were brought to Earth by meteor impacts or other means. To study meteorites, NASA and the National Science Foundation run the programme *ANSMET* (Antarctic Search for Meteorites) to collect meteorites in Antarctica in a sterile way for analysis. Due to the curvature of the Antarctic ice shield, meteorites move

Fig. 6.23 Living stromatolites in the *Hamelin Pool Marine Nature Reserve* in Shark Bay, Western Australia. Living stromatolites are rare and nowadays only exist in few local niches on Earth. They survived in the *Hamelin Pool* due to its exceptionally high salt content that is hostile for other life forms (Image: B. Ganse)

towards the Transantarctic Mountains and remain lying there. More than 20,000 meteorites have been found there and analysed since the 1970s. Similar programmes are run by other nations. NASA often sends astronauts to collect meteorites with the ANSMET programme to let them work and camp (in tents) in extreme conditions in isolation. None of these meteorites has so far indicated the existence of extraterrestrial life.

The *Drake equation*, also called *SETI equation* (*Search For Extraterrestrial Intelligence*) or *Green Bank formula*, is named after the American astronomer and astrophysicist Frank Drake, who developed it at the Green Bank observatory and presented it at a conference in 1961. It serves to estimate the number of civilisations of intelligent life forms in our galaxy that would be visible to radio telescopes. It does so by relating the following factors with each other: The average rate of star formation in the galaxy, the fraction of these stars being orbited by planets, the percentage of planets inside a star's habitable zone, the chance of such a planet actually producing intelligent life (speculative), the probability that this life would be willing to communicate (even more speculative) and the average lifespan of a civilisation in years. For many of the involved values, only coarse estimates are available.

The *Fermi-Paradox*, named after the Italian physicist Enrico Fermi, is a consequence of the Drake equation. If some best guesses for parameters of the Drake equation are inserted, one ends up with the prediction that our galaxy should be full of intelligent life and we should have made contact with them long ago. The fact that we have not met or seen a single one raises the question: why? Which of the parameters in the Drake equation is grossly misestimated? Is there a rule that civilisations reaching a certain technological level inevitably self-destruct, for example, by destroying their living environment or due to genocidal wars? Do they cloak to hide? Or is the development of life such an unlikely process that we are, in fact, the only ones?

Exactly this is the paradox. The Fermi-Paradox is a great topic for evening discussions around the campfire!

Crunching the Numbers: Drake Equation

The Drake equation gives an estimate (called N) for the number of intelligent civilisations in the Milky Way, whose signals can be received on Earth:

$$N = R_* \cdot f_p \cdot n_e \cdot f_\ell \cdot f_i \cdot f_c \cdot L$$

Where R_* is the rate of star formation in this galaxy (according to current estimates, this is about 7 stars per year), f_p is the percentage of stars orbited by one or more planets (data from the Kepler space telescope imply that this value is very close to 100%). All other quantities in this equation are completely unknown:

- n_e is the average number of planets in the habitable zone around a star. Unknown.
- f_ℓ is the probability that a planet that can support life will actually produce any. This number is somewhere in the range between 0 and 100%, but the precise value is completely unclear.
- f_i is the probability that life eventually develops intelligence. This number has been a topic of much discussion over many decades, but no certain value can be named, apart from the value range between 0 and 100%.
- f_c follows in the same vein and means the probability that an intelligent civilisation sends radio signals into space. Could there be other ways to communicate making radio waves irrelevant? Or is it essential for a species' survival to keep as quiet as possible?
- L is the time in years that such a civilisation would on average thrive and transmit signals. Maybe all intelligent civilisations perish after only a few centuries? Or do they continue existing for millennia once they reach a certain level of technological sophistication? Without any factual data, this number is pure speculation.

(continued)

Observations of the nearby universe have so far only yielded signs of a single intelligent civilization: humankind itself. For N, the value must therefore be one or larger, but by how much?

A common way to use the Drake equation is to assume that humanity is the only one ($N = 1$). Inserting purely speculative numbers and seeing what the equation predicts, quite disconcerting results can be achieved. If, for example, the assumption is inserted that every planet, which can support life, will eventually produce an intelligent species that communicates using radio waves ($f_\ell = f_i = f_c = 100\%$), then N should be phenomenally large...Unless the lifetime L of these civilisations is extremely short. Or the other way around: if the probability for sending radio signals f_c and the civilisation lifetime L are both large, then the only way to explain that our galaxy is not teeming with aliens is for the other numbers to be tiny, meaning that something really extraordinary must have happened on Earth to cause humankind to emerge (Fig. 6.24).

Already 2500 years ago, the Greek philosophers speculated on the theory of *panspermia* or *exogenesis*. It assumes life came to Earth from somewhere else, and that it could also spread from Earth to other places. Nobel Prize winner Francis Crick, together with Leslie Orgel, proposed the theory that extremely far progressed life forms could spread life to planets like Earth (directed

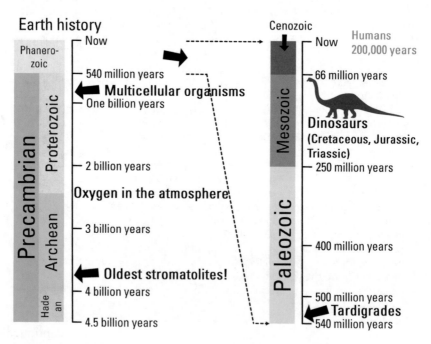

Fig. 6.24 Development of life on Earth. The time span of human existence is remarkably short

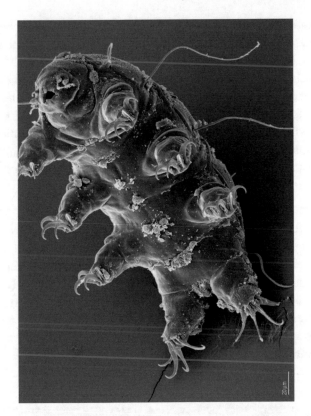

Fig. 6.25 Electron microscopy image of a tardigrade (Image: NASA)

panspermia). So far, there is no evidence supporting this theory. Experiments have shown that not only single-cell organisms, but also microscopic animals such as *tardigrades* can survive exposure to the vacuum of space. Tardigrades have eight legs and a body size of <1 mm (Fig. 6.25). Some such organisms are capable of *cryptobiosis*, putting their bodies into a dormant or suspended state, in which they can survive extreme environments (temperatures, pressures, chemicals, radiation, etc.) for a long period of time. In their dehydrated form, tardigrades have been shown to survive in the vacuum of space, exposed to cosmic radiation, for over 10 days. The African non-biting midge *Polypedilum vanderplanki* normally lives in temporary mud pools. In its larval stage, it can dry out completely, entering an anhydrobiotic state with remarkable resilience to temperature, radiation and vacuum conditions. Dried larvae of this insect species have survived exposure to space in the EXPOSE facility for over 400 days.

In the year 2010, animals were discovered that can live without free oxygen: three types of *Loricifera* (Danovaro et al. 2010). These animals live in salt-enriched water without oxygen in the Mediterranean Sea. Overall, much more life exists in extreme niches than previously assumed. The assumption that oxygen is a requirement for animals has been falsified! In addition, research results from studies dealing with life forms exposed to the direct space environment lead to the theory that they may survive while contained in rock and soil formations, protected from radiation, for millions of years. Altogether, there is growing evidence for the impression that life is much more resistant than previously assumed. In the context of these current research results, it becomes easy to imagine how life could spread in the universe while hidden in rocks.

> **Hint for Aspiring Spacefarers** Alien life forms would most probably not have eyes, legs and a body size like humans, elephants, rabbits or ants. Spacefarers should be prepared for microscopic life. It could simply show up as a slimey layer or microbial mat on a surface (Fig. 6.26). A microscope, tools and skills for staining, analysing and freezing the samples are helpful in this situation. Biologists as part of the team or direct contact to specialists on Earth might be desirable, too. At the same time, it is essential to not contaminate other worlds and samples.

Fig. 6.26 Thermophiles live at a hot spring in Aachen, Germany. They have adapted to the sulphur-rich hot water. Photosynthesis by algae seems to take place at the side (green colour). When finding life on a planet or moon, it could be a slime film. In case something like this is encountered during pace exploration, it is important to immediately take sterile samples for analysis (Images: B. Ganse)

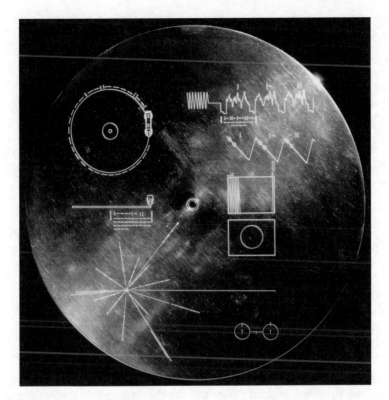

Fig. 6.27 Golden aluminium-cover of the Voyager-Golden-Records, titled *Sounds of Earth* (Voyager 1 and 2). The illustrations shown are a manual how to play the record. In 1977, LP records were state of the art technology (Image: NASA)

To get in contact with other civilisations, various types of messages have been sent in many ways. Metal plates with messages have customarily been placed on space probes, such as the *Voyager-Golden-Records* (Fig. 6.27) and the *Pioneer 10* and *11* messages (Fig. 6.28). In addition, many attempts have been made to send out radio signals, including the Arecibo message (Fig. 6.29). Answers, however, have not been received.

Nothing is known about the methods or codes alien life forms would use to communicate, but it makes sense to look for regular patterns or structures in radio transmissions. In 1977, the *Wow! signal* was recorded by a radio telescope. It came from the direction of the constellation Sagittarius and lasted for 72 s.

Despite intense attempts to capture it again and analyse it further, this was never achieved. Many possible sources of the signal are under discussion, including a pulsar (quickly rotating neutron star). Further radio signals that might originate from alien life have never been found. At the same time,

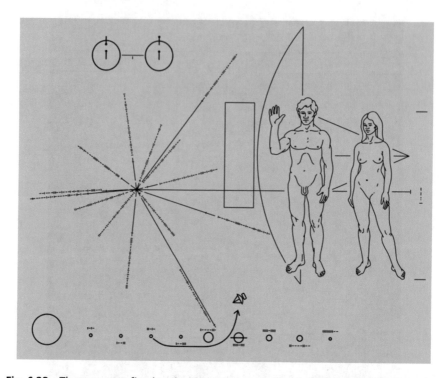

Fig. 6.28 The messages fixed to the Pioneer space probes included several illustrations. On top, the hyperfine structure transition of the hydrogen molecule. Below, the position of the Sun in relation to the centre of the Milky Way and several pulsars shown. At the bottom, the solar system and the trajectory of the probe are depicted. On the right, a man and a woman with the space probe in the background (Image: NASA)

signals were repeatedly sent from Earth not only towards the source of the Wow! signal, but also in the direction of Gliese 581, where the first Earth-like exoplanet has been detected. Some of these signals had a frequency close to the *hydrogen line* (1420 MHz), a resonance of the hydrogen atom. The assumption is that potential other civilisations, just as humans, use these wave lengths to map interstellar gas density. If they did, they could find these signals easily. Further plans include the installation of geometrical shapes in the Earth orbit or on the Moon as a sign that life exists here. It is possible that the artificial light of cities at night, with its sodium and mercury spectral lines, is visible from a far distance. Critical voices have raised concerns about drawing the attention of aliens towards us that may be superior and not come in peace.

When visiting other bodies such as planets, asteroids or moons, it is important to take care of planetary protection. This means trying to avoid contamination of the object with Earth microorganisms. NASA employs a

Fig. 6.29 The Arecibo message, a binary radio signal, was sent hoping that aliens would detect it. The first row shows the numbers 1–10. Below in purple the numbers 1, 6, 7, 8 and 15 are the ordinal numbers of the most important chemical elements for DNA. Shown below are the nucleotides of human DNA, the structure of the DNA double helix (blue) and the number of nucleotides as a white bar. The body size of a human shown in red is encoded on its left. On the right, the number of humans on Earth is shown. The solar system and an illustration of the radio transmitter (Arecibo observatory) are also depicted. The question is, whether aliens would be able to decode and understand it (Image: NASA)

person whose job it is to take care of this important issue: the *Planetary Protection Officer*. It will be interesting to see what comes first: Life from Earth being introduced to other planets, or alien life forms discovered on other worlds?

References

Danovaro, R., DellAnno, A., Pusceddu, A., Gambi, C., Heiner, I., & Kristensen, R. M. (2010). The first metazoa living in permanently anoxic conditions. *BMC Biology, 8,* 30.

Huygens, C. (1698). *Kosmotheoros, de terris cœlestibus, earumque ornatu, conjecture at constantinum hugenium*. Den Haag: Adriaan Moetjens.

Kant, I. (1755). *Allgemeine Naturgeschichte und Theorie des Himmels*. Hamburg: Verlag tredition. ISBN 978-384-241518-8.

Zekl, H. G., & Copernicus, N. (2006). Das neue Weltbild: Sonderausgabe der Philosophischen Bibliothek. ISBN 978-3-78731-800-1.

Conclusion

This concludes the Spacefarer's Handbook. An attentive reader should now be equipped with all basic knowledge for a journey into space and back. The authors wish best of luck in implementing further ideas and plans, conquest and settlement of the solar system or just simply dreaming in the comfort at home.

Hint for Aspiring Spacefarers Finally, a couple of points of practical advice:

- No matter how bad the smell inside the spacecraft may get: Always keep the windows closed!
- Do not try to land on the surface of the Sun.
- There is no beer volcano on Enceladus and any rumours about it are entirely unfounded.
- It is inadvisable to stick one's head into a rocket engine's exhaust.
- Even though the name would imply otherwise, Proxima Centauri is really not close to us and not a suitable holiday destination.

© Springer-Verlag GmbH Germany, part of Springer Nature 2020
B. Ganse, U. Ganse, *The Spacefarer's Handbook*, Springer Praxis Books,
https://doi.org/10.1007/978-3-662-61702-1

Index

© Springer-Verlag GmbH Germany, part of Springer Nature 2020
B. Ganse, U. Ganse, *The Spacefarer's Handbook*, Springer Praxis Books,
https://doi.org/10.1007/978-3-662-61702-1

Printed in the United States
By Bookmasters